Ariacutty Jayendran

Mechanical Engineering

Grundlagen des Maschinenbaus in englischer Sprache

T0225077

Ariacutty Jayendran

Mechanical Engineering

Grundlagen des Maschinenbaus in englischer Sprache

Mit 168 Abbildungen sowie einer englisch-deutschen und deutsch-englischen Vokabelübersicht

Teubner

Bibliografische Information Der Deutschen Bibliothek
Die Deutsche Bibliothek verzeichnet diese Publikation in der Deutschen Nationalbibliografie;
detaillierte bibliografische Daten sind im Internet über <http://dnb.d-nb.de> abrufbar.

Prof. Dr. Ariacutty Jayendran, M. Sc. Ph. D. London, Chartered Engineer/ MIEE London. Jetzt emeritiert, zuvor Prof. der Physik an den Universitäten Khartoum, Sudan und Colombo, Sri Lanka.

1. Auflage September 2006

Der B.G. Teubner Verlag ist ein Unternehmen von Springer Science+Business Media.
www.teubner.de

Umschlaggestaltung: Ulrike Weigel, www.CorporateDesignGroup.de
Druck und buchbinderische Verarbeitung: Strauss Offsetdruck, Mörlenbach
Gedruckt auf säurefreiem und chlorfrei gebleichtem Papier.

ISBN 978-3-8351-0134-0

Vorwort

Dieses Buch richtet sich an Studierende der Ingenieurwissenschaften und praktisch tätige Ingenieure, die sich im Verstehen englischsprachiger Lehrbücher und Fachtexte profilieren und ihr sprachliches Verständnis verbessern wollen. Es enthält das Grundwissen wichtiger Fachgebiete des Maschinenbaus und möchte eine Brücke schlagen zum Einstieg in englische Fachliteratur.

Das Buch ist als kompaktes Handbuch konzipiert und verbindet theoretische und praktische Lehrinhalte. Die inhaltliche Gliederung und Kapitelabfolge entspricht anderer deutscher oder englischer Handbücher dieses Themenkreises.

Mit diesem Handbuch wird den Studierenden die Möglichkeit gegeben, ein sprachliches Grundwissen in technischem Englisch zu erwerben und gleichzeitig inhaltliche Grundkenntnisse der einzelnen Fachgebiete des Handbuchs kompakt vorzufinden. Es ist in gut verständlichem Englisch verfaßt. An zahlreichen Stellen finden sich nach englischen Schlüsselbegriffen die deutschen Entsprechungen in Klammern beigefügt. So ist dem Text sprachlich sehr gut zu folgen, das englische Vokabular wird zunehmend verständlich und der Einstieg in andere englische Texte auf diese Weise sehr erleichtert.

Das Buch ist so kompakt wie möglich verfaßt worden und enthält die Grundkenntnisse einzelner Bereiche wie Mechanik, Maschinenelemente, Thermodynamik oder auch Fertigungstechnik in der didaktisch üblichen Reihenfolge. Der Schwerpunkt liegt nicht auf dem Unterricht der englischen Sprache, sondern auf der Vermittlung von Grundkenntnissen einzelner Bereiche des Maschinenbaus und Ingenieurwesens auf der Basis der englischen Sprache. Es kann von Studierenden und Ingenieuren als Referenzbuch genutzt werden. Die Zeichnungen sind nach der "British Standard Specification" erstellt, Symbole entsprechen denen in englischer Fach- und Lehrbuchliteratur. Die Leser erhalten so einen Einblick in die Unterschiede der Normung und Formelnotation zwischen deutscher und englischer Literatur. Ein Formelverzeichnis, eine englisch-deutsche und deutsch-englische Vokabelliste und ein sowohl deutsches als auch englisches Stichwortverzeichnis unterstützen dies.

Zum weiterführenden Verstehen englischer Texte des Maschineningenieur-wesens empfehle ich elektronische Medien wie das "Cambridge Advanced Learner's Dictionary on CD-ROM" (Cambridge University Press) zu nutzen. Hier finden sich englisches und amerikanisches Fachvokabular mit Audiounterstützung.

Ausgezeichnete Hilfe bieten auch Wörterbücher im Internet, wobei besonders die Wörterbuchsammlung unter http://dict.leo.org zu empfehlen ist.

Mein besonderer Dank gilt meiner Frau Christel, ohne deren Hilfe ich dieses Buch nicht hätte schreiben können.

Wetter, im Juli 2006   Ariacutty Jayendran

Contents

Inhaltsverzeichnis

8   Contents

I Mechanics
(Mechanik)

1 Statics (Statik)

1.1 Forces (Kräfte)

A force is a physical quantity which causes a change in *the motion of a body* or in *its form*. A force has both *magnitude* and *direction,* and is therefore classified as a *vector quantity*. When *specifying a force*, it is not enough to specify the magnitude and direction of the force. It is also necessary to specify *where a force acts*, and this is done by specifying its *line of action*. An external force may be applied at *any point along its line of action* without changing its effect on the body.

A force is usually *represented in a diagram* by an arrow as shown in Fig 1.1. The length of the arrow is proportional to the *magnitude* of the force, while the direction of the arrow is the same as the *direction* of the force.

A number of forces F_1, F_2.........,F_n can be added together to form a *single resultant force* F_r which has the same effect on the body as all the individual forces acting together. The SI unit of force is called the Newton.

The subject of *statics* is mainly concerned with the action of *external forces* which are necessary and sufficient to keep rigid bodies in a *state of equilibrium*. *Changes in form* do not come within the *scope of statics.*

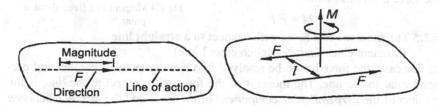

| Fig 1.1 Diagram of a force | Fig 1.2 Diagram of a couple |

1.2 Couples and moments (Kräftepaare und Momente)

1.2.1 Couples. (Kräftepaare)

A couple consists of *two equal and opposite parallel forces* F separated from each other by a *distance l*. A couple tends to *rotate a body* and its effect cannot be reduced to that of a single force. The moment of a couple M is a measure of its ability to cause rotation and is given by the expression

$$\text{Moment } M = \text{Force } F \times \text{distance of separation } l$$

The *sense of rotation* of a couple can be *clockwise* or *counterclockwise.*

1.2.2 Displacement of a couple (Verschiebung eines Kräftepaares)

The moment of a couple remains the same even if the forces forming the couple are *displaced* and their *magnitude and direction changed* (Fig 1.3), provided the forces *remain parallel* to each other, have *the same moment*, and *remain in their original plane* (or in a parallel plane). The sense of rotation must also remain the same.

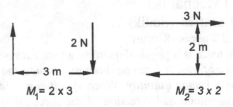

$M_1 = 2 \times 3$ $M_2 = 3 \times 2$

Fig 1.3 Two equivalent couples with forces which are different in magnitude and direction

1.2.3 Vector representation of a couple
(Vektor Darstellung eines Kräftepaares)

The *magnitude, direction* and *sense* of rotation of a couple can be completely represented by a *single moment vector M* as shown in Fig 1.2. This is a line drawn perpendicular to the plane in which the couple acts and whose length is proportional to the moment of the couple.

1.2.4 The moment of a force about a point
(Moment einer Kraft bezüglich eines Punktes)

The moment of a force about a point P is the product of the *magnitude of the force* and the *perpendicular distance* of the line of action of the force from this point.

$$M = F\,l$$

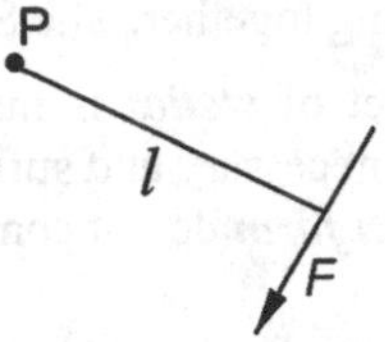

Fig 1.4 Moment of a force about a point

1.2.5 The moment of a force with respect to a straight line
(Moment einer Kraft bezüglich einer Linie)

In this case, the force has to be resolved into two components parallel and perpendicular to the line. The moment of the force with respect to the line, is the product of the *perpendicular component* of the force and the *distance between the line of action* of the force and the *given straight line*.

1.2.6 Moving a force to act through any arbitrary point
(Verschiebung einer Kraft)

If we have a force F acting through *any point* A in a body (Fig 1.5(a)), we can show that this is equivalent to *the same force F* acting through an *arbitrarily chosen point* B and *a couple* (as shown in Fig 1.5(c)).

To show this, we place two equal and opposite forces F at point B (Fig 1.5 (b)). This should have *no effect* on the body. It can be seen that the original force F at A and the force $-F$ at B form a couple having a moment $M = Fd$.

Thus as shown in Fig 1.5 (c), we have replaced the original force F at A by the same force acting through an arbitrary point B plus a couple of moment $M = Fd$.

It is easy to see (by reversing the procedure) that *the reverse is also true*. This means that *a force F and a couple M acting on a body are **equivalent to a single force F** acting through another point in the body*.

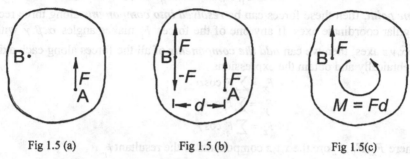

Fig 1.5 (a)   Fig 1.5 (b)   Fig 1.5(c)

Fig 1.5 Moving a force to act through any arbitrary point

1.3 Composition and resolution of forces
(Zusammensetzen und Zerlegen von Kräften)

1.3.1 Parallelogram and triangle of forces
(Kräfteparallelogramm und Kräftedreieck)

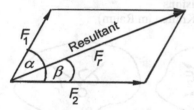
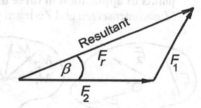

Fig 1.6 (a) Parallelogram of forces   Fig 1.6 (b) Triangle of forces

If we have two forces F_1 and F_2 acting at a point as shown in Fig 1.6(a), these two can be added vectorially to give a *single resultant force F_r* which is the *diagonal* of the *parallelogram* formed by the forces. It is often more convenient to use a *triangle* which is half the *parallelogram* (Fig 1.6(b)) to find the resultant graphically. *Adding forces vectorially* in this way is called the *composition of forces*.

Here the resultant is $F_r = \sqrt{F_1^2 + F_2^2 + 2F_1F_2\cos\alpha}$ and $\beta = \arcsin\dfrac{F_1\sin\alpha}{F_r}$

Conversely, a single force F_r can be *resolved* (or split) into two forces F_1 and F_2 in *any two directions*, where

$$F_1 = \frac{F_r\sin\beta}{\sin\alpha} \quad \text{and} \quad F_2 = F_r\cos\beta - F_1\cos\alpha$$

1.3.2 Composition and resolution of forces acting on a rigid body at a point

(Zusammensetzen und Zerlegen von Kräften mit gemeinsamem Angriffspunkt)

If we have a number of forces $F_1, F_2, \ldots, F_n$ acting on a rigid body *at the same point*, then these forces can be *resolved into components* along three rectangular coordinate axes. If any one of the forces F_i makes angles α, β, γ with the x,y,z axes, then we can *add the components* of all the forces along each axis algebraically and obtain the expressions

$$F_x = \sum F_i \cos\alpha$$
$$F_y = \sum F_i \cos\beta$$
$$F_z = \sum F_i \cos\gamma$$

where F_x, F_y, F_z are the x,y,z components of the resultant F_r.

Therefore
$$F_r = \sqrt{F_x^2 + F_y^2 + F_z^2}$$

The angles made by the resultant with the three axes are given by

$$\alpha_r = \arccos\frac{F_x}{F_r}, \quad \beta_r = \arccos\frac{F_y}{F_r}, \quad \gamma_r = \arccos\frac{F_z}{F_r}$$

1.3.3 Composition and resolution of a number of forces with different points of application in three dimensions

(Zusammensetzen und Zerlegen von Kräften im Raum)

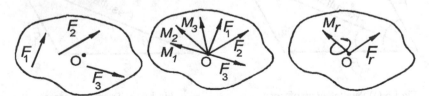

Fig 1.7 (a)   Fig 1.7(b)   Fig 1.7 (c)

Fig 1.7 (a) Three forces acting at different points in a body.
Fig 1.7 (b) An arrangement equivalent to (a) with the same three forces
         acting at an arbitrary point O together with three couples.
Fig 1.7 (c) Equivalent arrangement with a single force and a single couple.

Consider a body acted on by three forces F_1, F_2, F_3 *each acting at a different point* as shown in Fig 1.7 (a). As shown in section 1.2.7 each of these forces can be moved individually to act through an *arbitrary point* O provided we also *introduce a couple* for each force transferred. The result of such a transfer is shown in Fig 1.7 (b) which shows the same three forces acting at an arbitrary point O together with three vectors which represent the three couples which have been added. This may be further simplified as shown in Fig 1.7 (c) to a single force F_r and a single couple of moment M_r. The point O is arbitrary, but the *magnitude* and *direction* of the force F_r will always *be the same*.

The value of F_r is obtained in the same way as has been shown in section 1.3.2. and is given by

$$F_r = \sqrt{F_x^2 + F_y^2 + F_z^2}$$

The values of M_1, M_2, M_3 and M_r will however *depend on the point chosen*.

The magnitude of M_r may be found by a similar process to that for finding F_r.

It is given by

$$M_r = \sqrt{M_x^2 + M_y^2 + M_z^2}$$

The moment vector M_r makes angles $\alpha_r, \beta_r, \gamma_r$ with the x,y,z axes where

$$\alpha_r = \arccos\frac{M_x}{M_r}, \ \beta_r = \arccos\frac{M_y}{M_r}, \ \gamma_r = \arccos\frac{M_z}{M_r}$$

1.3.4 Conditions for equilibrium (Gleichgewichstbedingungen)

(a) When a rigid body is acted on by a number of forces at *different points* and in *different directions*, equilibrium exists only when $F_r = 0$ and $M_r = 0$ which means that the following *six conditions* must be satisfied.

$$F_x = 0, F_y = 0, F_z = 0 \text{ and } M_x = 0, M_y = 0, M_z = 0$$

(b) When all the forces are *in the same plane*, equilibrium exists when the following *three conditions* are satisfied.

$$F_x = 0, F_y = 0, M_z = 0$$

1.3.5 Graphical methods (Graphische Methoden)

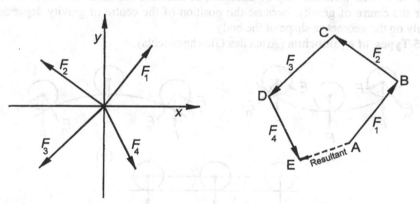

Fig 1.8 (a) Shows four forces acting at a point   Fig 1.8 (b) Polygon of forces

Consider four forces which are in the same plane and acting at a point as shown in Fig 1.8 (a). The resultant of these forces can be found by *repeated use* of the *triangle of forces* (Fig 1.8 (b)). If we *omit the closing sides* of the triangles, we have a *polygon of forces,* which in this case is *not closed* (i.e. has an open side EA). The sides representing the forces can be taken in any order, the arrows (showing their direction) go round the *periphery* of the polygon in the *same direction*. The magnitude and direction of the *resultant* is represented by the *clos-*

ing side of the polygon EA. Its *direction* however is *counter* to that of the other sides of the polygon. *Equilibrium exists* when the *polygon closes* and the *resultant is zero*. If the *forces are not in the same plane*, the resultant may be found by adding these forces in space, forming in effect a *polygon in space*.

1.4 Centre of gravity, centre of mass and the centroid

(Schwerpunkt, Massenmittelpunkt und Fläschenschwerpunkt)

Any body can be supposed to be composed of a small number of particles each

of mass dm. The gravitational force acting on each of these is gdm. All these *small forces* together will have a *resultant* which is equal to the *total force of gravity* mg acting on the body whose total mass is m. The force will act through a point G called the *centre of gravity* (Fig 1.9). The total mass of the body may be considered to be concentrated at the point G. The *centre of gravity* is therefore also termed the *centre of mass.*

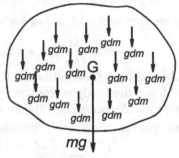

Fig 1.9 Centre of gravity

When the *density* of the body is *uniform* through the whole of the body, then bodies having the *same geometric shape* will have the same centre of gravity, although their densities may be different. In this case the term *centroid* is used for the centre of gravity, because the position of the centre of gravity depends only on the geometric shape of the body.

1.5 Types of equilibrium (Arten des Gleichgewichts)

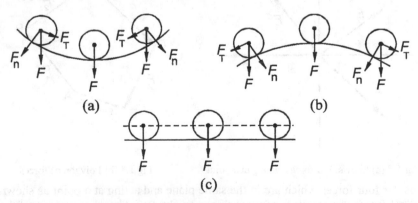

Fig 1.10 (a) Stable equilibrium (b) Unstable equilibrium (c) Neutral equilibrium

Consider a body which has the freedom to move but which remains at rest. It is said to be in a *state of equilibrium*. We distinguish between *three different types of equilibrium.*

(a) **Stable equilibrium** – An example of this is a ball rolling on a concave surface as shown in Fig 1.10 (a). A state of stable equilibrium exists when the slightest movement of the body *raises its centre of gravity*. When it moves even slightly away from the minimum position, the force F has a *tangential component* F_T which *moves it back* to its *original minimum position*.

(b) **Unstable equilibrium** – A state of unstable equilibrium exists, when the movement of the body *lowers its centre of gravity* as in the case of a ball rolling on a convex surface (Fig1.10 (b)). In this case the tangential component F_T tends to *move it away* from the *equilibrium position*.

(c) **Neutral equilibrium** - Neutral equilibrium exists, when a ball rolls on a flat surface as shown in Fig 1.10 (c). Here any movement away from the equilibrium position *does not create a force* which tends to move it towards or away from its original position.

1.6 Friction (Reibung)

1.6.1 Introduction (Einführung)

The term friction refers to the *force of resistance* which arises, when two surfaces which are in contact with each other, *slide* or *tend to slide* against each other. When the surfaces are *dry* and *free of contamination* by liquids, the resistance is called *dry friction*. An example of this is the friction that exists between surfaces of a *brake shoe* and a *brake drum*. This friction is *absolutely necessary* for the functioning of the brake, and must be maintained at a high level. For many other applications however, it is desirable to reduce the friction by *lubrication*. The term *lubrication* refers to the maintenance of a *thin film of fluid or gas* between the sliding surfaces.

1.6.2 Static friction (Haftreibung)

Consider a body with a flat lower surface lying on a flat horizontal surface. Normally the weight F_G acts vertically downwards, and the body remains in *equilibrium* because F_G is opposed by a *vertical force of reaction* F_N. If we now apply a horizontal force F_T which tends to move the body, the body does not move because the force F_T is *opposed* by a *frictional force* F_R. If we gradually *increase* F_T, the body remains at rest until the force F_R reaches a *limiting value* F_{RO}.

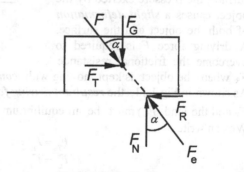

Fig 1.11 Static friction

If F_T *exceeds* this value, then *motion takes place*. We can see that the total force F acting on the body is the vector sum of F_G and F_T and is inclined to the vertical. The force of reaction F_e is the vector sum of F_N and F_R and acts along the *same line of action* as F but in the *opposite direction*. Under equilibrium

conditions
$$F_G = F_N = F\cos\alpha$$
$$F_T = F_R = F\sin\alpha$$
$$\frac{F_R}{F_N} = \tan\alpha$$

If $\alpha = \rho_0$, when F_R has its maximum value F_{R0}, then

Coefficient of static friction $\mu_0 = \dfrac{F_{R0}}{F_N} = \tan\rho_0$

Angle of static friction $\qquad \rho_0 = \arctan\mu_0$

1.6.3 Sliding friction (Gleitreibung)

When the body starts moving (or sliding) on the surface, the frictional force F_R opposing the motion is *less than* the *static value* F_{R0}. We can write

Coefficient of sliding friction $\mu = \dfrac{F_R}{F_N}$ and angle of friction $\rho = \arctan\mu$

The value of this coefficient depends on *a number of factors* such as the nature of the two surfaces, the lubrication, the velocity of the motion, etc.

1.6.4 Rolling resistance (Rollreibung)

Rolling is used as a way of moving bodies *in preference to sliding*. Frictional resistance to rolling is *considerably smaller* than that due to *sliding*. When a cylindrical or spherical object rests on a flat surface, the pressure exerted by the object causes a *slight deformation* of both the object and the surface. A driving force F is required to overcome the frictional resistance

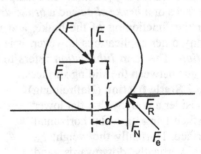

Fig 1.12 Rolling resistance

F_R when the object is kept moving with *constant velocity* along the surface. As shown in Fig 1.11, the *resultant driving force F* (which is the vector sum of F_T and the load F_L) must be in equilibrium with the *force of reaction* F_e. We can write

$$F_T = F_R$$
$$F_L = F_N$$

Also by equating moments $\qquad F_T r = F_L d$

By substituting we have $\qquad F_R r = F_N d$

$$\frac{F_R}{F_N} = \frac{d}{r}$$

From this we see that the *coefficient of rolling friction* is

$$\mu_r = \frac{d}{r}$$

To ensure that *only rolling* occurs and that *no sliding* takes place, it is necessary that
$$\mu_r < \mu_0$$

1.6.5 Resistance to motion (Fahrwiderstand)

When considering the resistance to motion of a vehicle, it is necessary to take into account the *resistance of the bearings* whose effect must be added to the frictional resistance. The effect of both these resistances can be combined to give a *combined coefficient* of friction μ_f. In this case the following condition has to be satisfied.
$$\mu_f < \mu_0$$

1.6.6 Friction in screws (Reibung beim Schrauben)
(a) Screw with square threads

(Schraube mit Rechteckgewinde)
The *tightening* or *loosening* of a screw corresponds to the *up* or *down* movement of a load on an *inclined plane*. If

$p =$ pitch of the screw (see page 118)

$\alpha =$ inclination of the thread to the horizontal (see page 118)

$r_2 =$ mean radius of the screw

$\mu = \tan \rho =$ coefficient of friction of the screw thread

$$\tan \alpha = \frac{p}{2\pi r_2}$$

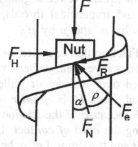

Fig 1.13 Friction in screw threads

A force F_H acting at right angles to the axis of the screw is required to keep the screw in uniform motion.
$$F_H = F \tan(\alpha \pm \rho)$$

The moment of the force required to tighten or loosen the screw is given by
$$M = F_H r_2 = F \tan(\alpha \pm \rho) r_2$$

In the absence of friction ($\rho = 0$), the force would be
$$F_0 = F \tan \alpha$$

From this we can state that the efficiency of the screw is
$$\eta = \frac{F_0}{F_H} = \frac{F \tan \alpha}{F \tan(\alpha + \rho)} = \frac{\tan \alpha}{\tan(\alpha + \rho)} \quad \text{when the nut is raised}$$

and
$$\eta = \frac{\tan(\alpha - \rho)}{\tan \alpha} \quad \text{when the nut is lowered}$$

When $\alpha \leq \rho_0$, the moment M is *negative* or *zero*. When $\alpha = \rho_0$, the efficiency is
$$\eta = \frac{\tan \alpha}{\tan 2\alpha} \approx 0.5$$

(b) Screws with V and trapezoidal threads
(Schraube mit Spitz- und Trapezgewinde)

As seen from Fig 1.14 the force F_N acting at right angles to the surface of the thread is given by

$$F_N = F / \cos(\beta/2)$$

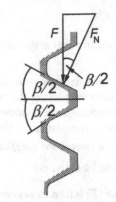

The friction in the thread is greater than for a rectangular thread and is given by

$$\mu F_N = \frac{\mu F}{\cos(\beta/2)}$$

We can write
$$\mu' = \tan \rho' = \frac{\mu}{\cos(\beta/2)}$$

From this it is clear that that the relationships given for square (or rectangular) threads hold also for V and trapezoidal threads, provided we replace ρ by ρ' and μ by μ'.

Fig 1.14 V and trapezoidal threads

1.6.7 Friction due to ropes, belts, etc. (Seilreibung)
Consider a rope, belt, or band which is stretched over a drum or pulley. The tensional force F_1 is **greater** than the force F_2 because of the frictional force F_R along the **area of contact** between the belt and the drum. Let α be the angle subtended by the arc of contact between belt and drum in radians.

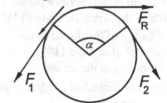

Fig 1.15 Friction due to ropes and belts

$$F_1 = F_2\, e^{\mu\alpha}$$

$$F_R = F_1 - F_2 = F_1\left(\frac{e^{\mu\alpha}-1}{e^{\mu\alpha}}\right)$$

The sliding coefficient of friction μ has to be used in the above relationship when slip takes place, and the static coefficient μ_0 when there is no slip.

1.6.8 Friction in pulleys (Reibung in Rollen)
(a) Fixed pulley (Feste Rolle)

The frictional resistance that exists between the pulley and its bearing, together with the resistance of the rope to flexing, necessitates a pulling force (or effort) F which is **larger** than the load F_1 . If the load and the pulling force both move through a distance s, then the efficiency of the fixed pulley is

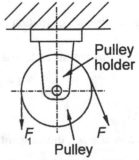

Pulley holder

Pulley

$$\eta_f = \frac{F_1 s}{F s} = \frac{F_1}{F}$$

Fig 1.16 Fixed pulley

The efficiency is about 0.96 for plain bearings and 0.97- 0.98 for ball bearings.

(b) Single moving pulley (Lose Rolle)

The load is divided between two ropes. Movement of pulling force $s_f = 2 \times$ movement of load s_1.

Also

$$F_1 = F + F_0$$

$$\eta_m = \frac{F_1 s_1}{F s_f} = \frac{(F + F_0) s_1}{2 F s_1}$$

$$\eta_m = \frac{F + F \eta_f}{2F} = \frac{1 + \eta_f}{2}$$

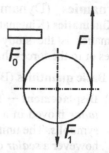

Fig 1.17 Moving pulley

The pulling force $F = \dfrac{F_1}{\eta_f + 1}$. If $\eta_f = 0.95$, then $\eta_m = (1 + 0.95)/2 = 0.975$.

It follows that a *moving pulley* has a *higher efficiency* than a *fixed pulley*.

1.6.9 Friction in pulley systems

Pulley systems which consist of more than two pulleys have a *mechanical advantage* and enable a *small force* (or effort) F to lift a *much bigger load* F_1.

Mechanical advantage = Load / Effort = F_1 / F

If the total number of pulleys (without counting the last direction changing pulley) is n, then the number of sections of rope carrying the load is $(n+1)$.

Therefore

$$s_f = (n+1) s_1$$

If there are no frictional losses $F = \dfrac{F_1}{n+1}$. If the efficiency of the pulley system

is η_r, then the lifting force is $F = \dfrac{F_1}{\eta_r (n+1)}$ when the direction changing pulley is

a moving pulley. If the turn-around pulley is a fixed pulley, then $F = \dfrac{F_1}{\eta_r \eta_f (n+1)}$.

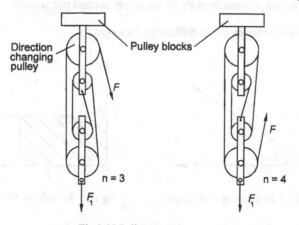

Fig 1.18 Pulley systems

2 Dynamics (Dynamik)

2.1 Kinematics (Kinematik)

Kinematics is the study of the motion of bodies without reference to their masses or to the forces that cause their motion.

2.1.1 Basic quantities (Basisgrößen)

a) **Displacement** — The displacement of a point may be defined as the *distance* moved in a *specified direction*. It is a *vector* quantity and has the symbol *s*. The unit of displacement is the metre (m). The term *distance* is however a *scalar* quantity with the *same unit* i.e., metre .

b) **Velocity** — The *velocity* of a point is also a *vector* quantity, and is the *rate of change* of *displacement* with time.

$$\text{Velocity} \quad v = \frac{ds}{dt}$$

The unit of velocity is the metre per second (m/s).

c) **Speed** — The *speed* of a point is a *scalar* quantity and refers to the *rate of change* of *distance* with time. When a body moves with *constant speed,* the *magnitude* is *constant* while the *direction* may be *changing*. A body moving round a circle is moving with *constant speed* but *not* with *constant velocity*. Speed has the same units as velocity (m/s).

d) **Acceleration** — The *acceleration* of a point is a *vector* quantity and is defined as the *rate of change* of *velocity*.

$$\text{Acceleration} \quad a = \frac{dv}{dt}$$

A body can move with *uniform* or with *nonuniform* acceleration. When a body moves with *negative acceleration*, meaning that its velocity is decreasing with time, it undergoes *deceleration*. Acceleration and deceleration are expressed in units of metre per second per second (m/s^2).

2.1.2 Uniform motion (Gleichförmige Bewegung)

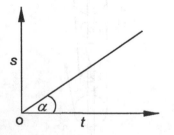   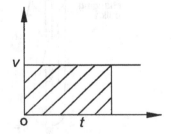

Fig 1.19 Distance vs time diagram   Fig 1.20 Velocity vs time diagram

A body is said to be in a state of **uniform motion** if its *velocity is constant*, which implies that its **acceleration is zero**. The change in distance with time can be shown in a s,t diagram as shown in Fig 1.19. The change in velocity with time can be shown in a v,t diagram as shown in Fig 1.20. In the case of uniform motion the velocity is constant and this is represented in the diagram as a horizontal line. The *shaded part* under the velocity line represents the *distance travelled* during a given time.

2.1.3 Nonuniform motion (acceleration or deceleration)

(Ungleichförmige Bewegung - Beschleunigung oder Verzögerung))
When a body is in a state of **nonuniform motion** then its *velocity is changing*, and it travels different distances during equal intervals of time. The change of distance with time is shown in the s,t diagram of Fig 1.21. The slope of the curve at any point gives the velocity at that instant of time.

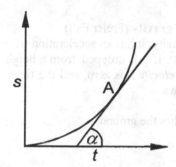

Fig 1.21 Nonuniform motion

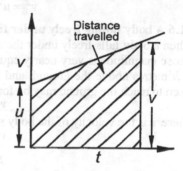

Fig 1.22 Uniform acceleration

2.1.4 Uniform acceleration (Gleichmäßig beschleunigte Bewegung)

The v,t diagram in Fig 1.22 shows the change in velocity with time for a body which has an initial velocity u and a uniform (or constant) acceleration a. If the body moves for a time t, then

$$\text{Increase in velocity} = at$$
$$\text{Final velocity } v = \text{ initial velocity } u + \text{ increase in velocity } at$$
$$\text{or} \quad v = u + at$$

2.1.5 Distance travelled when a body moves with uniform acceleration

(Weglänge unter gleichmäßig beschleunigter Bewegung)
The distance travelled can be found as shown below.

$$\text{Average velocity} = \frac{u+v}{2}$$
$$\text{Since} \quad v = u + at$$
$$\text{Average velocity} = \frac{u+u+at}{2} = u + \frac{1}{2}at$$
$$\text{Distance travelled} = \text{average velocity} \times \text{time}$$

$$\text{Distance travelled} = (u + \frac{1}{2}at)t$$

$$\text{or} \qquad s = ut + \frac{1}{2}at^2$$

The *distance travelled* is easily shown to be equal to the *area* of the *trapezoidal space* below the velocity line (shown shaded in Fig 1.22).

A useful *third equation* may be obtained which *does not involve the time* t.

If we square both sides of the equation

$$v = u + at$$
$$v^2 = u^2 + 2uat + a^2t^2$$

$$\text{we obtain} \qquad v^2 = u^2 + 2a(ut + \frac{1}{2}at^2)$$

The term in the bracket is equal to s, and therefor

$$v^2 = u^2 + 2as$$

2.1.6 A body falling freely under the action of gravity (Freier Fall)

When a body falls freely under the action of gravity, it has an acceleration of g whose magnitude is very nearly equal to 10 m/s^2. If it is dropped from a height of h metres above the ground, and the *initial velocity* u is zero, and the time t taken to reach the ground may be found as follows.

$$v = u + gt$$

where v is the velocity of the body when it reaches the ground.

$$v = 0 + gt$$

$$t = \frac{v}{g}$$

Time taken to reach the ground is given by

$$\text{Since} \quad v^2 = 0 + 2gh \qquad t = \frac{v}{g} = \frac{\sqrt{2gh}}{g} = \sqrt{\frac{2h}{g}}$$

2.1.7 A body thrown vertically upwards (Senkrechter Wurf)

Let the *initial vertical upward velocity* be u. In this case the body will suffer a deceleration g until its *velocity becomes zero*. After that it will *move downwards* with an acceleration g. For the upward movement, we can use the equation

$$v^2 = u^2 - 2gh$$

$$0 = u^2 - 2gh$$

The maximum height reached is $\qquad h = \dfrac{u^2}{2g}$

To find the time taken to reach this height, we write

$$v = u - gt$$

$$0 = u - gt$$

$$t = \frac{u}{g}$$

2.1.8 A body projected with an initial horizontal velocity
   (Horizontaler Wurf)

Consider a body initially projected with a horizontal velocity u_h. The motion of

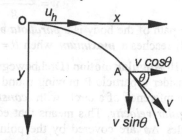

Fig 1.23 A body projected horizontally

the body is a *combination* of the *vertical motion* of the falling body and the *horizontal motion* due to the velocity of projection. The horizontal velocity remains constant. The path of the motion is a parabola. If we resolve the velocity v at a point A along its path, then

$$v_x = v\cos\theta = u_h = \text{const.}$$

$$v_y = v\sin\theta = gt$$

$$v = \sqrt{v_x^2 + v_y^2} = \sqrt{u_h^2 + (gt)^2}$$

Velocity after falling through a distance h is

$$v = \sqrt{u_h^2 + 2gh}$$

2.1.9 A body projected upwards at an oblique angle(Wurf schräg nach oben)

Consider a body projected with an initial velocity u at an angle θ to the horizontal. We can resolve u into *two components* and as was stated in the last section, the *horizontal component* $u\cos\theta$ remains *constant* right through the motion. The vertical component $u\sin\theta$ undergoes a deceleration g until the body rises to a maximum height h. It then moves downwards with an acceleration g.

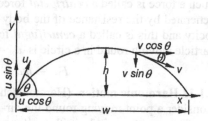

Fig 1.24 Body projected at an oblique angle

1. Time taken to reach the maximum height h is $\qquad t = \dfrac{u\sin\theta}{g}$

2. Time required for the body to reach the ground is $\qquad 2t = \dfrac{2u\sin\theta}{g}$

3. Maximum height reached by the body is $\qquad h = \dfrac{u^2\sin^2\theta}{2g}$

4. Horizontal distance travelled by body is given by $\quad w = u\cos\theta \times 2t$

$$w = u\cos\theta \times \frac{2u\sin\theta}{g}$$

$$w = \frac{u^2 \sin 2\theta}{g}$$

The path of the body is a **parabola** and the **horizontal distance** travelled by the body reaches a **maximum** when $\theta = 45°$.

2.1.10 Angular motion (Drehbewegung)

Consider a particle P moving round the circumference of a circle with **constant angular velocity**. This means that equal angles $\Delta\phi$ are covered by the point in equal intervals of time Δt. It follows that equal distances Δs will also be traced in equal intervals of time Δt. Angular velocity $\omega = \frac{\Delta\phi}{\Delta t}$ and $\Delta\phi = \frac{\Delta s}{r}$

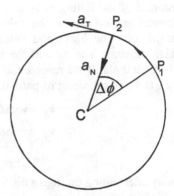

$$\omega = \frac{1}{r}\frac{\Delta s}{\Delta t} = \frac{v}{r} \ (v = \text{linear velocity of P})$$

$$v = r\omega$$

Fig 1.25 Particle moving round a circle

The particle P moves with constant angular velocity and constant speed. However its **direction** and **velocity** are **changing** and it is therefore undergoing acceleration. A force has to be applied constantly, to deviate it towards the centre. Such a force is called a **centripetal** force. An equal and opposite inertial force is generated by the resistance of the body to the change in the direction of the velocity and this is called a **centrifugal** force. The acceleration required to keep a particle moving round in a circle is $a_N = v^2/r$. Therefore the force required is

$$F_N = ma = mv^2/r$$

2.1.11 Harmonic motion (Harmonische Bewegung)

Consider a point moving round the circumference of a circle with uniform angular velocity ω in an anticlockwise direction. As the point P moves round the circumference, the **projection of P** on any diameter XX′ (which is the point A) moves in a **straight line**. If the point starts at X′ and moves along the circumference to P in time t, then the angle $\theta = \omega t$. If $CP = r$, then

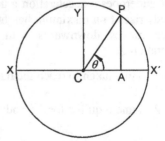

Fig 1.26 Harmonic motion

1. Distance moved by A in time t is $\quad s = X′A = r - r\cos\omega t$

2. Velocity of A along the x-axis is $\quad \dfrac{ds}{dt} = \omega r \sin\omega t$

3. Acceleration of A along the x-axis is $\dfrac{dv}{dt} = \omega^2 \cos\omega t$

4. The period $\qquad\qquad\qquad\qquad T = \dfrac{2\pi}{\omega}$

The *period T* is the time taken by the point P to complete one cycle of motion (or go round the circle once). This is also the time taken for the point A to complete a full cycle on the *x*-axis by moving from X′ to X and back to X′. The point X goes through a reciprocating motion. The *displacement, velocity* and *acceleration* of the point A can be seen to be *sinusoidal functions* of time. The reciprocating motion of the point A is called *harmonic motion*.

2.2 Kinetics of translational motion
(Kinetik des translatorisch bewegten Körpers)
The term kinetics refers to the study of the *motion of bodies* as a consequence of the *forces* and *moments* which act on them.
2.2.1 Work (Mechanische Arbeit)
Work (or mechanical work) is said to be done when the point of application of the force moves. The amount of work done is measured by the *product of the force* and the *distance moved* in the direction of the force.

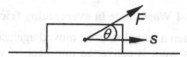

Fig 1.27 Work done by a force

Work done = Force x distance moved in the direction of the force.

Work is a *scalar quantity*.

If we consider the body shown in Fig 1.27 to be moved through a horizontal distance *ds*, the work done by the force is given by

$$dW = F\cos\theta\, ds$$

$$W = \int_{s_1}^{s_2} F\cos\theta\, ds$$

The work done by a couple of moment *M* when it causes a body to rotate through an angle *dθ* is given by

$$dW = M\, d\theta$$

$$W = \int_{\theta_1}^{\theta_2} M\, d\theta$$

The SI unit of work is the Joule (J) and is the work done when the point of application of a force of one Newton (N) moves through a distance of one metre (m) in the direction of the force.

Joule = Newton x metre

$$1\,J = 1\,Nm = 1\,kg\,m^2/s^2$$

2.2.2 Work done in lifting a weight (Arbeit der Gewichtskraft)

The weight or gravitational force acting on a mass m is equal to mg. This can be assumed to be a constant. If a body is raised through a vertical distance h, the work done is

$$W = mgh$$

2.2.3 Work done in stretching a spring (Formänderungsarbeit einer Feder)

When a spring is stretched, the restoring force is proportional to the elongation of the spring when it is stretched.

$$\text{Restoring force} = \text{constant} \times \text{elongation}$$

$$|F| = cs$$

$$\text{Work done} = \int_{s_1}^{s_2} csds = \frac{c(s_2^2 - s_1^2)}{2}$$

2.2.4 Work done in overcoming friction (Reibungsarbeit)

When a body has to be moved against frictional forces, work has to be done and this work is converted into heat. If we have a body of weight mg resting on a horizontal plane, then the force of friction is given by μmg where μ is the *coefficient of friction*. If we now apply a horizontal force which moves the body through a distance s, then

$$\text{Work done} = \mu mgs$$

If the body rests on an *inclined plane* of slope α and if we apply a force parallel to the slope to move it through a distance s,

$$\text{Work done} = \mu mgs \cos \alpha$$

2.2.5 Power (Leistung)

Power may be defined as the *rate of doing work*.

$$P(t) = \frac{dW}{dt}$$

$$\text{and} \qquad W = \int P(t)dt$$

The SI unit of power is the watt (W) and corresponds to work being done at the rate of one joule per second.

$$1W = 1J/s$$

Larger units are the kilowatt (kW) and the megawatt (MW).

2.2.6 Efficiency (Wirkungsgrad)

The efficiency of a machine (or a process) is the *ratio of the useful work* done by the machine (or process), to the *total work* put into the machine (or process). This ratio is usually expressed as a percentage and we may write

$$\text{Efficiency } \eta = \frac{\text{Work output}}{\text{Work input}} \times 100\%$$

The efficiency is always less than 100%.

2.2.7 Newton´s laws of motion (Newtonsche Grundgesetze)

When a single force or a number of forces act on a body, the body moves. The relationships between the *forces which cause the motion* and the *motion* itself, are clearly stated in Newton´s laws of motion.

1) The first law states that a body continues in its *state of rest* or of *uniform motion in a straight line* unless it is compelled by some external force to act otherwise. The tendency of a body to remain in a state of rest or continue to move in a straight line is described as *inertia.*

2) The second law states that the *rate of change of momentum* of a body is proportional to the *applied force* and takes place in the direction in which the force acts. The momentum p of a body is defined as the *product* of *its mass* and *its velocity*. Momentum is a *vector* quantity and has the units kg m/s.

$$\text{Momentum} = \text{mass} \times \text{velocity}$$
$$p = mv$$

According to Newton´s second law

$$F = \frac{d}{dt}(mv)$$

$$F = m\frac{dv}{dt} = ma$$

3) Newton´s third law of motion states that when a *force acts on a body*, an *equal* and *opposite force* acts on another body. This has been sometimes stated in the form, for *every action* there is an equal *and opposite reaction*. An example of this is a bullet fired from a gun. When the bullet is fired, equal and opposite forces act on the bullet and the gun during the time that the bullet is passing through the gun.

2.2.8 Energy and the law of conservation of energy
(Energie und Energie Erhaltungssatz)

The term energy is used to denote the *ability* of a body or a system *to do work*. Anything that can do work is said to *possess energy*. Energy like work is a *scalar quantity*. There are many types of energy like mechanical energy, heat energy, electrical energy, chemical energy, etc.

There are also many types of *mechanical energy*. Among these are *kinetic energy*, *potential energy* and the *deformation energy* of an elastic body. One form of energy can be converted into the other.

A swinging pendulum is an example of a body whose energy can be kinetic or potential or a mixture of both. It is completely potential at the beginning of the swing and completely kinetic when passing through its rest position.

This is an example of the *law of conservation of energy* which states that energy cannot be destroyed. It can only converted to another form of energy.

2.2.9 Conversion of work into translational kinetic energy
(Umwandlung von Beschleunigungsarbeit in kinetische Energie)

A moving body has the *energy of motion* or *kinetic energy*. When a force acts on a body and the body moves, work is done by the force. This work is converted into kinetic energy. The kinetic energy of a body of mass m moving with a velocity v has been defined by the expression

$$\text{Kinetic energy} = \frac{1}{2}mv^2$$

$$\text{Work done} = \int_{s_1}^{s_2} F ds = \int_{s_1}^{s_2} m\frac{dv}{dt} ds = \int_{v_1}^{v_2} mv dv$$

$$= \frac{1}{2}mv_2^2 - \frac{1}{2}mv_1^2 = E_2 - E_1$$

$$\text{Work done} = \text{Change in kinetic energy}$$

2.2.10 Impulse and momentum (Kraftstoß und Impuls)

The *product of force* and *time* has been termed *impulse*. This is a *vector quantity* with units of Newton second (Ns). Impulse is useful in calculating the effect of a force *on a body* for *a short time*, and also when the force *is not constant*.

$$F = ma = m\frac{dv}{dt}$$

$$\text{Impulse} = \int_{t_1}^{t_2} F dt = \int_{u}^{v} m dv = m(v - u)$$

where u is the initial velocity and v the final velocity

The *momentum* of a body mv is defined as the product of its mass and its velocity. It has units of kg m/s.

It follows that   Impulse = Change in momentum

2.2.11 Law of conservation of linear momentum (Impulserhaltungssatz)

The law of conservation of momentum states that the *linear momentum* of a system of bodies *is unchanged* if there is *no external force* acting on the system.   If

$$\sum \int_{t_1}^{t_2} F_i dt = \sum m_i(v_i - u_i) = 0$$

then

$$\sum m_i v_i = \sum m_i u_i$$

It follows that **momentum is conserved** when no external force acts on the system. A simple example is a system consisting of **two bodies** which collide with each other. We can write $m_1u_1 + m_2u_2 = m_1v_1 + m_2v_2$

where u_1, u_2 are the initial and v_1, v_2 the final velocities. It follows that

Momentum before collision = Momentum after collision

2.3 Kinetics of rotational motion (Kinetik der Rotation des starren Körpers)

2.3.1 Rotation of a rigid body about a fixed axis

(Rotation eines starren Körpers um eine feste Achse)

Consider the motion of a small particle of mass dm which is part of a rigid body rotating about an axis passing through O (Fig 1.28). If the distance of the particle from the axis is r and the tangential force acting on it is dF_T, then

$$dF_T = a_T dm$$

where a_T is the tangential acceleration of the particle. The moment of this force is

$$dM = r \, dF_T = r a_T dm$$

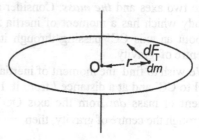

Fig 1.28 Rotation about an axis

The sum of the moments of all the elementary forces dF_T acting on the total mass m is given by

$$M = \int r a_T dm$$

We know that $a_T = r \dfrac{d\omega}{dt}$ where ω is the angular velocity. Substituting for a_T in the previous equation, we have

$$M = \int r a_T dm = \int r \, r \frac{d\omega}{dt} dm$$

The angular acceleration ω is the same for all the particles dm and so we can write $M = \dfrac{d\omega}{dt} \int r^2 dm$

We can write $M = J\dfrac{d\omega}{dt}$ where $J = \int r^2 dm$ is known as the **moment of inertia**.

2.3.2 Moment of inertia J (Trägheitsmoment)

The **moment of inertia** depends both on the **mass** of the body and the **distribution of mass** in the body. It also depends on the **axis of rotation**. This is a **scalar** quantity and its units are kg m². The equation $M = J\dfrac{d\omega}{dt}$ in rotational motion is similar to $F = ma$ in linear motion.

We can consider force and moment to be similar quantities in the sense that *force* causes *linear motion* while *moment* causes *rotational motion*. Also *moment of inertia* in *rotational motion* is similar to *mass* in *linear motion*.

2.3.3 The parallel axes theorem (Satz von Steiner)

The moment of inertia of a body about any given axis is equal to the moment of inertia about a *parallel axis* through the *centre of gravity*, plus the *product* of the square *of the distance* between the two axes and *the mass*. Consider a body which has a moment of inertia J about an axis CC′ passing through its centre of gravity.

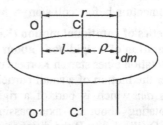

Fig 1.29 Parallel axis theorem

We wish to find the moment of inertia J' about another axis OO′ which is parallel to CC′ and at a distance l from it. If r is the perpendicular distance of an element of mass dm from the axis OO′, and ρ its distance from the axis CC′ through the centre of gravity, then

$$J' = \int r^2 dm = \int (l+\rho)^2 dm$$

$$J' = \int l^2 dm + \int 2l\rho\, dm + \int \rho^2 dm$$

$$J' = l^2 \int dm + 2l \int \rho\, dm + \int \rho^2 dm$$

The integral $\int dm = m$. The integral $\int \rho\, dm = 0$ because this is the sum of the moments of all the elements of mass about the axis CC′(see p58). Since this passes through the centre of gravity, it follows that the integral must be zero. The integral $\int \rho^2 dm = J$. Therefore we can write

$$J' = l^2 m + 0 + J$$

or $$J' = J + ml^2$$

If a body is *composed of a number of parts*, the moment of inertia of the body J about any axis is equal to the sum of the moments of inertia of the parts J_1, J_2,........, J_n about the same axis. Therefore

$$J = J_1 + J_2 + \ldots\ldots\ldots\ldots + J_n$$

2.3.4 The radius of gyration (Trägheitsradius)

The mass of a body is usually distributed over a large volume. We can however imagine the entire mass to be *concentrated at a point*. The distance k (from the axis of rotation) at which the point mass has to be placed so that the moment of inertia remains the same is called the radius of gyration.

$$J = k^2 m$$

$$k = \sqrt{\frac{J}{m}}$$

2.3.5 Rotational kinetic energy (Rotationsenergie)

The definition of kinetic energy as given by $E = \frac{1}{2}mv^2$ also holds for each particle of mass dm in a rigid rotating body. If we use the relation

$$v = r\omega$$

Rotational energy $E_{rot} = \int \frac{1}{2} dm v^2 = \frac{1}{2} \int \omega^2 r^2 dm$

Since ω is a constant for all the particles in the body

$$E_{rot} = \frac{1}{2} \omega^2 \int r^2 dm$$

$$E_{rot} = \frac{1}{2} J \omega^2$$

Here again it can be seen that J has a similar position in *rotational energy* to that which m has in *translational energy*.

2.3.6 Work done in increasing the rotational energy of a body
(Umwandlung von Beschleunigungsarbeit in Rotationsenergie)

When a *couple* acts on a body that can be rotated, the *angular velocity* of the body is *increased*. When the angular velocity is increased, the *rotational kinetic energy* is also *increased*.

Consider a tangential force F_T acting on a body and rotating it through a small angle $d\theta$.

Linear distance moved $ds = r d\theta$

Work done $dW = F_T ds = F_T r d\theta$

Since $F_T r = M$

Therefore $dW = M d\theta$

$$W = \int M d\theta$$

Since $M = J \frac{d\omega}{dt}$

$$W = \int J \frac{d\omega}{dt} d\theta$$

$$W = J \int_{\omega_1}^{\omega_2} \omega \, d\omega$$

$$W = \frac{1}{2} J \left(\omega_2^2 - \omega_1^2 \right)$$

$$W = E_2 - E_1$$

Work done = Change in rotational energy

A moving body can have both translational and rotational kinetic energy.

Total kinetic energy = Translational kinetic energy + Rotational kinetic energy

$$E = \frac{1}{2}mv^2 + \frac{1}{2}J\omega^2$$

2.3.7 Angular Impulse (Drehimpuls)

Analagous to the quantity *linear impulse* in *linear motion*, we can also define the quantity

$$\text{Angular impulse} = Mdt = J\frac{d\omega}{dt}dt = Jd\omega$$

If the angular velocity of a body changes from ω_1 to ω_2 in time t_1 to t_2, then

$$M(t_2 - t_1) = J(\omega_2 - \omega_1)$$

2.3.8 Conservation of angular momentum (Drehimpuls Erhaltungssatz)

It can be shown that similar to the case of the conservation of linear momentum, the *angular momentum* of a system of bodies *is unchanged* if there are *no external forces* (*or couples*) acting on the system.

2.4 Impact (Stoß)

2.4.1 Forces acting during an impact (Kräfte beim Stoß)

The term *impact* applies to the collision between two bodies in which *relatively large forces* act between them over a comparatively *short period of time*. The motion of the bodies is changed by the impact. *No external forces* act on the bodies *during an impact* and therefore the *forces* exerted by the bodies on each other must *be equal and opposite* as would be expected from Newton´s third law of motion. It follows that the law of conservation of momentum holds and that the *total momentum after the collision* is equal to the *total momentum before the collision*. Therefore

$$m_1 u_1 + m_2 u_2 = m_1 v_1 + m_2 v_2$$

where u_1, u_2 are the velocities of m_1, m_2 before the impact

and v_1, v_2 are the velocities of m_1, m_2 after the impact.

2.4.2 Collinear or direct central impact (Gerader zentrischer Stoß)

When two bodies collide with each other, a *tangential plane* drawn through the *point of contact* of the two bodies is called the *plane of contact*. A line drawn *normal* to the *plane of contact* through the *point of contact*, is called the *line of impact*.

If the line of impact passes through the *centres of gravity* of both bodies, then the impact is called a *central impact*. Any other impact is called an *eccentric impact*.

If the linear momentum vectors of the bodies are also *directed along the line of impact* at the beginning of the impact, then the impact is called a *collinear impact* or a *direct central impact*. Any other impact is called an *oblique impact*.

2.4.3 Elastic and inelastic impacts (Elastischer und unelastischer Stoß)

During a collision the bodies undergo *deformation*. If the bodies are restored to their original state without suffering any *permanent deformation*, the impact is *elastic* and there is *no loss of energy* during the impact. This is the case when two billiard balls or two rubber balls strike each other. On the other hand, if there is permanent deformation and the *restoration of energy is incomplete*, the impact is *inelastic*.

2.4.4 Elastic collinear impact (Elastischer gerader zentrischer Stoß)

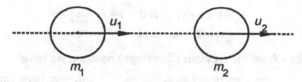

Fig 1.30 (a) Approaching spheres before impact

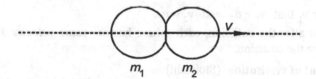

Fig 1.30 (b) Spheres during contact

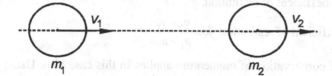

Fig 1.30 (c) Spheres moving apart after impact

Consider two spheres of mass m_1, m_2 moving along the same line with velocities u_1, u_2 where $u_1 > u_2$ (Fig 1.30 (a)). The impact can be divided into two stages.

First (or compression) stage

This stage begins with the moment of contact, and ends when the distance between the centres of gravity is a minimum (Fig 1.30(b)). At the end of this stage they have a common velocity v.

Applying the law of conservation of momentum, we have

$$m_1 u_1 + m_2 u_2 = m_1 v + m_2 v$$

$$v = \frac{m_1 u_1 + m_2 u_2}{m_1 + m_2}$$

Second (or restoration) stage
This stage begins when the *separation* between the bodies is a *minimum* and *ends with* the *complete end of contact* between them. The velocity of the first body is changed from v to v_1, and that of the second from v to v_2. In the second or restoration stage, the same force acts on both spheres as in the first compression stage. Consequently each sphere undergoes *equal changes in velocity* in the two stages.

$$u_1 - v = v - v_1 \quad \text{and} \quad v - u_2 = v_2 - v$$
$$v_1 = 2v - u_1 \quad \text{and} \quad v_2 = 2v - u_2$$

Substituting for v from the previous (first stage) equation we have

$$v_1 = \frac{(m_1 - m_2)u_1 + 2m_2 u_2}{(m_1 + m_2)} \quad \text{and} \quad v_2 = \frac{(m_2 - m_1)u_2 + 2m_1 u_1}{(m_1 + m_2)}$$

If the spheres have the same mass, then $m_1 = m_2$, we can show by substituting in the equation for v_1 that $v_1 = u_2$ and $v_2 = u_1$.
This means that after the collision, each sphere has *the same velocity* that **the other had** before the collision.

2.4.5 Coefficient of restitution (Stoßzahl)
Real bodies are neither fully elastic nor fully inelastic. In the case of real bodies, the ratio of *the velocity of separation* to the *velocity of approach* is a *constant* called the coefficient of restitution.

$$\text{Coefficient of restitution} \quad k = \frac{v_2 - v_1}{u_1 - u_2}$$

The law of conservation of momentum applies in this case also. Using this fact and the above equation for k, we can show that

$$v_1 = \frac{m_1 u_1 + m_2 u_2 - m_2 (u_1 - u_2)k}{m_1 + m_2}$$
$$v_2 = \frac{m_1 u_1 + m_2 u_2 + m_1 (u_1 - u_2)k}{m_1 + m_2}$$

The energy loss ΔE during the collision can be shown to be

$$\Delta E = \frac{1}{2} \frac{m_1 m_2 (u_1 - u_2)^2 (1 - k^2)}{m_1 + m_2}$$

3 Hydrostatics (Hydrostatik)

3.1 Properties of fluids and gases
(Eigenschaften der Flüssigkeiten und Gase)

An *ideal fluid* (as compared to a solid) is only able to transmit *normal forces* and is unable to resist *shear forces*. In practice all real fluids show some *internal friction* and *shear resistance*. Fluids can be divided into two types, liquids (which are virtually *incompressible*) (and gases which are *compressible*).

3.1.1 Density (Dichte)
The density of a *fluid* is defined as its mass per unit volume. The density of a *liquid* changes with *temperature*, but shows *negligible change* with *pressure*. The density of a *gas* is however a function of both *temperature* and *pressure*.

3.2 Hydrostatic pressure (Hydrostatischer Druck)
The force exerted by a liquid on a surface is always *normal* to the surface. The pressure exerted on a surface is the force per unit area and the unit of pressure is Newton per square metre (N/m^2). This unit has a special name, the Pascal (Pa).

$$\text{Pressure} \quad P = F_N/A$$

where F_N is the force acting normal to the surface and A the area of the surface.

$$1 \text{ Pa} = 1 N/m^2$$
$$1 \text{ bar} = 10^5 \text{ Pa} = 0.1 MPa$$

3.3 The transmission of pressure (Druck Ausbreitungsgesetz)
When the pressure due to the force of gravity is neglected, the hydrostatic pressure is transmitted through a liquid in accordance with Pascal's principle. According to this principle, the pressure exerted by a fluid (which is in equilibrium) on a surface is always at *right angles* to the *surface*. Moreover, the pressure at any point in the fluid has the *same magnitude* in *all directions*.

If an *external pressure* is exerted on a closed volume of a liquid (for example by a piston), this pressure is transmitted *unchanged* to *all parts of the fluid* and in *all directions*.

3.4 Applications of the law of pressure transmission
(Anwendungen des Druck Ausbreitungsgesetzes)

3.4.1 Force exerted on the wall of a vessel
(Wanddruckkraft)

If we consider the pressure exerted on the *hemispherical surface* BCD shown in Fig 1.31, then the total force exerted on the hemispherical surface is the product of the pressure P and the *projection of the hemispherical surface* on a plane at right angles to the

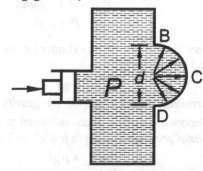

Fig 1.31 Force on a curved surface

direction of the force. Here the

force is $PA = \dfrac{P\pi d^2}{4}$

3.4.2 Cylindrical tube (Rohr)

If we have a cylindrical tube of diameter d and length l, *the projection* of *half the cylinderical surface* on the *diametral plane* is

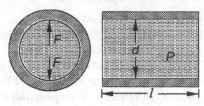

Fig 1.32 Force on one half of a cylinder

$$A = dl$$

Then the force exerted on *each half* of the tube surface is given by

$$F = PA = Pd$$

The forces tend to *tear the tube apart*.

3.5 Pressure due to the weight of a fluid

(Druckverteilung durch Gewichtskraft der Flüssigkeit)

Consider the forces acting on a *cylindrical part* of a *fluid* which is in *equilibrium*. If the surface area of each flat surface of the cylinder is A, then the volume of the cylinder is

$$V = hA$$

and the weight of the cylinder of liquid is given by

$$F_G = hA\rho g$$

If the forces acting on the flat surfaces are F_1 and F_2 as shown in the figure, then for equilibrium

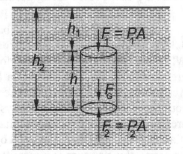

Fig 1.33 Pressure at a depth h

$$F_2 = F_1 + F_G$$
$$P_2 A = P_1 A + \rho g h A$$
$$P_2 = P_1 + \rho g h$$

If the upper flat surface of the cylinder lies on the *top surface* of the liquid, then $P_1 = 0$ and we can write

$$P = \rho g h$$

The hydrostatic pressure caused by gravity is therefore *proportional to the depth* h below the surface. If an additional pressure P_a is exerted on the surface, then the *total pressure* at a depth h is given by

$$P = P_a + \rho g h$$

This total pressure is everywhere the same at this depth and also on the wall of the container.

3.6 Hydrostatic forces exerted on the walls of open containers
(Hydrostatische Kräfte gegen ebene Wände offener Gefäße)

3.6.1 Force exerted on the base of a vessel (Bodenkraft)

The force exerted on the *plane base* of an *open vessel* (as shown in Fig 1.34) is

$$F_b = \rho g h A$$

and is *independent* of the *shape* of the vessel.

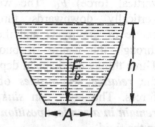

Fig 1.34 Force on the base of a vessel

3.6.2 Force exerted on the sidewalls (Seitenkraft)

Consider a vessel whose sidewall is *inclined at an angle* α to the horizontal. The force F which acts on an area A of the sidewall is proportional to

1. the depth h_C of the centre of gravity of the area A
2. the density of the liquid ρ
3. the area A of the sidewall which is being considered

$$F = \rho g h_C A$$

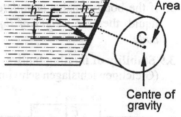

Area

Centre of gravity

Fig 1.35 Force exerted on a sidewall

The *location of the force* can be shown to be at a depth of

$$h_F = h_C + I_G (\sin\alpha)^2 / h_C A$$

where I_G is the areal moment of inertia taken round its centre of gravity.

3.7 Buoyancy and the principle of Archimedes
(Auftrieb und das Prinzip von Archimedes)

Archimedes´principle states that when a body is *partially or completely immersed* in a fluid, an *upthrust* or *buoyant vertical force* F_U equal to the *weight of liquid displaced* acts on the body (through the centre of buoyancy B).

Upthrust $F_U = \rho g V_D$ = weight of liquid displaced

Where V_D is the *volume of fluid displaced*. If the body is made of material of density ρ' then, the weight of the body $F_G = \rho' g V_D$

The *apparent weight* of the cylinder F_A is equal to the difference between the real weight and the upthrust.

Apparent weight = Real weight – Upthrust

$$F_A = (\rho' - \rho) g V_D$$

3.8 Floating bodies (Schwimmende Körper)

When a body is placed in a liquid, there are two forces acting on it, the weight F_G, and the upthrust (or buoyant vertical force) F_U. The weight acts vertically downwards through the centre of gravity G and the upthrust through the centre of buoyancy B. Three things can happen depending on the *relative magnitudes* of the two forces. The body can *sink, float* or *remain in a random position* anywhere in the fluid.

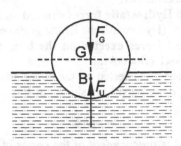

Fig 1.36 A floating body

1. A body *sinks* when its *weight is greater* than the *upthrust*.
2. If a floating body is still *partially immersed* in the liquid, then the *weight of liquid displaced* is exactly equal to the *weight of the body*. The centre of buoyancy is below the centre of gravity on the same vertical line.
3. If the body remains *completely immersed* in any random position in the liquid, then the *weight of liquid displaced* is equal to the *weight of the body*.

3.9 Stability of floating bodies
(Gleichgewichtslagen schwimmender Körper)

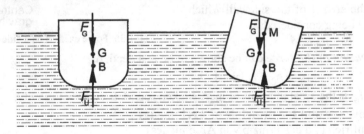

Fig 1.37(a) Body in a stable position   Fig 1.37 (b) Body in a displaced position

When a floating body is in a *stable position* as shown in Fig 1.37(a), then the centre of gravity and the centre of buoyancy lie on the same vertical line.

The body is in *stable equilibrium* if the *centre of buoyancy* B is *below* the centre of gravity G. If the body is *displaced from the stable position* as in Fig 1.37(b), a *restoring couple* is created, and this tends to return the body back to its original position. The point M where the line of action of the buoyant force intersects the centre line of the body is called the *metacentre*.

It can be easily shown that if the *centre of buoyancy* B is *above* the centre of gravity G, a couple is created which tends to *overturn* the body. In this case, the body is in a state of *unstable equilibrium*.

4 Fluid dynamics (Hydrodynamics)
(Dynamik der Flüssigkeiten, Hydrodynamik)
4.1 Basic concepts (Grundlagen)
4.1.1 Ideal and real fluids (Ideale und nichtideale Flüssigkeiten)

Substances which are initially at rest respond to the application of a *shear stress* in different ways. Consider two very large plates of equal size (as shown in Fig 1.38) separated by a small distance y. The upper plate is *moved* while the lower plate is kept *stationary*. The space between the plates is filled with a substance which may be a solid or a fluid.

The surfaces of the substance adhere to the plates in such a manner, that the upper surface of the substance moves at the *same velocity* as the upper plate while the lower surface remains *stationary*. If a constant force F_s is applied, the moving plate attains a constant velocity u. The shear stress that arises is defined by $\tau = \dfrac{F_s}{A_s}$ where A_s is the surface area of the plate. The applied stress causes the substance to be deformed and the rate of deformation is given by du/dy.

Deformation characteristics for various substances are shown in Fig 1.39.

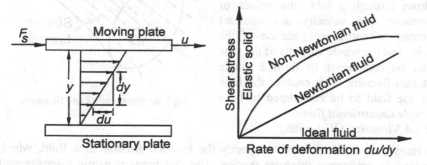

Fig 1.38 Flow of a substance between
parallel plates

Fig 1.39 Deformation characteristics of
different substances

An ideal (or elastic) solid *resists shear stress* and its rate of deformation will be zero. If a substance *cannot resist* even the *slightest shear stress* without flowing, then it is a fluid. An ideal fluid does not have internal friction and hence its deformation rate lies along the x-axis as shown in Fig 1.39. Real fluids have *internal friction* and their rate of deformation depends on the applied stress. If the deformation is directly proportional to the stress, it is called a *Newtonian fluid*. If the deformation is not directly proportional it is called a *non-Newtonian fluid*.

Fluids can be of two types, *compressible* and *incompressible*. A liquid can be considered to be incompressible, while gases and vapours are compressible.

4.1.2 Steady and unsteady flow (Stationäre und nichtstationäre Strömung)

The flow of a liquid is said to be *steady* if the fluid properties like pressure and density are only dependent on the *space coordinates* and *not on the time*. The flow is *unsteady* if the properties at a point *vary with time*.

4.1.3 Streamlines, stream tubes and filaments
    (Stromlinien, Stromröhre und Stromfaden)

If we follow the movement of a fluid particle through a fluid, then a *streamline* is a line which gives the *direction of the velocity* of the particle at each point in the stream. If there is steady flow, then the streamlines are identical to the *flow lines* of the liquid. They are *fixed in space* and do not change with time. If there is *unsteady flow*, the streamlines *change their position* with time.

If a number of streamlines (in steady flow) are connected by a *closed curve*, we have a *stream tube*. They form a boundary through which the fluid particles do not pass. Parts of a stream tube over which the pressure and the velocity *remain constant* are called *stream filaments*. When an ideal fluid flows through a tube, the values of pressure and velocity are constant across the entire cross-section of the tube, and the whole contents of the tube can be considered to form a single stream filament. This enables the flow of the fluid to be considered to be a *single dimensional flow.*

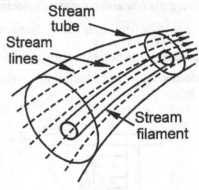

Fig 1.40 Stream tubes and filaments

4.1.4 Viscosity (Viskosität)

Viscosity is the term used to specify the *internal friction* of a fluid, which means its *resistance to shear motion*. The two terms dynamic viscosity and kinematic viscosity are defined as follows:

Dynamic viscosity $\quad \eta = \dfrac{\tau}{(du/dz)} = \dfrac{\text{Shear stress}}{\text{Rate of deformation}}$

Kinematic viscosity $\quad v = \dfrac{\eta}{\rho} \quad = \dfrac{\text{Dynamic viscosity}}{\text{Density}}$

Units: Dynamic viscosity

10 Poise $= 1 \text{Ns/m}^2$
    1 P $= 0.1 \text{ Ns/m}^2$

Units: Kinematic viscosity

10 Stokes (St) $= 1 \text{m}^2/\text{s}$
    1 St $= 10^{-4} \text{ m}^2/\text{s}$

4.1.5 Reynolds number (Reynoldssche Zahl)

The amount of influence that *frictional forces* have on the flow of a fluid is indicated by the Reynolds number which has the symbol Re (or R (American)).

$$Re = \frac{w \, d\rho}{\eta} = \frac{wd}{\nu}$$

where w is the average velocity, d the diameter of the tube, ρ the density, η the dynamic viscosity and ν the kinematic viscosity.

In *streamline* or *laminar flow* the particles of a fluid move in parallel layers, while in *turbulent flow* the particles have *additional velocity components* in the x, y, z directions.

Small values of the Reynolds number correspond to laminar flow while large numbers correspond to turbulent flow. The change from laminar to turbulent flow takes place when the critical value of 2300 is exceeded.

4.1.6 The Mach number (Machsche Zahl)

The Mach number is the ratio of the *average fluid velocity* w to the *speed of sound* c. The symbols used are Ma or M (American).

$$Ma = \frac{w}{c}$$

If w is small compared to c, the compressibility of the fluid does not play a role. The flow of gases can be considered to be incompressible when $Ma < 0.3$ corresponding to $w = 100$ m/s in the case of air.

4.2 The basic equations of fluid flow (Grundgleichungen der Strömung)

4.2.1 The continuity equation (Kontinuitätsgleichung)

The continuity equation is a special case of the law of *conservation of mass*. When a fluid flows through a tube, the mass of fluid flowing through any cross-section must be the same. If the cross-sections at two points in a tube are A_1 and A_2 and the correspond-, ing average velocities are w_1 and w_2, then the *mass of fluid* q_m flowing through any cross-section per second is

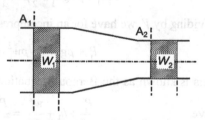

Fig 1.41 The continuity equation

$$q_m = A_1 w_1 \rho_1 = A_2 w_2 \rho_2$$

If the fluid is incompressible, the volume q_v flowing through must also be constant. Therefore

$$q_v = A_1 w_1 = A_2 w_2$$

4.2.2 Energy equation for fluid flow (Bernoulli equation)
(Energieerhaltungssatz der Strömung, Bernoullische Druckgleichung)

Application of the principle of *conservation of energy* to fluid flow results in the following equation which is known as the *Bernoulli equation*. Consider a tube containing a fluid as shown in Fig 1.42. The fluid is made to flow from positions 1 to 2 between heights h_1 and h_2.

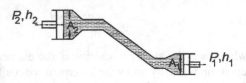

Fig 1.42 Energy equation for fluid flow

An *amount of work* P_1V_1 is done on the fluid at 1 and an amount of work P_2V_2 is done by the fluid at 2. Although the work may be supposed to be done by the movement of the pistons, their presence is not really necessary.

The fluid possesses at positions 1 and 2 potential energies of mgh_1 and mgh_2 and kinetic energies $\frac{1}{2}mw_1^2$ and $\frac{1}{2}mw_2^2$ where m is the mass and w the average velocity.

If friction losses are neglected, we can write

Energy at the end of the process	=	Energy at the beginning of the process	±	Amount of work delivered to the fluid

$$mgh_2+\frac{m}{2}w_2^2 \quad = \quad mgh_1+\frac{m}{2}w_1^2 \quad + \quad \left(P_1V_1-P_2V_2\right)$$

Since $m = \rho V$,

$$P_1V_1+\rho Vgh_1+\frac{1}{2}\rho Vw_1^2 = P_2V_2 +\rho Vgh_2+\frac{1}{2}\rho Vw_2^2 = \text{constant}$$

Dividing by V we have for an incompressible fluid

$$P_1 +\rho gh_1+\frac{1}{2}\rho w_1^2 = P_2 +\rho gh_2+\frac{1}{2}\rho w_2^2 = \text{constant}$$

This is known as the Bernoulli equation. If this equation is divided by ρg, we have

$$\frac{P_1}{\rho g}+h_1 +\frac{w_1^2}{2g} = \frac{P_2}{\rho g}+h_2 +\frac{w_2^2}{2g}$$

The *individual terms* correspond to *heights* (with units in metres (m)). For this reason, the height corresponding to each type of energy can be termed a *head*.

$$\frac{P}{\rho g} = \text{static pressure head of the fluid}$$

$$h = \text{potential head of the fluid}$$

$$\frac{w^2}{2g} = \text{velocity head of the fluid}$$

4.2.3 Individual pressure terms (Einzelne Druckglieder)

In cases where the *fluid flow is horizontal*, the gravitational term may be omitted in the Bernoulli equation and we can write

$$P_1 + \frac{1}{2}\rho w_1^2 = P_2 + \frac{1}{2}\rho w_2^2$$

The pressure P is called P_{stat}, and this is the *static pressure* which causes the flow. The term $\rho w^2/2$ is a measure of *the kinetic energy* per unit volume of the fluid flowing with an average velocity w. This is called the *dynamic pressure* P_{dyn}. The sum of these terms is equal to the *total pressure* P_{tot}.

$$P_{tot} = P_{stat} + P_{dyn}$$

4.3 Applications of the Bernoulli equation
(Anwendungen der Bernoulligleichung)

4.3.1 Fluid flow past an obstacle
(Auftreffen einer Strömung auf ein festes Hindernis)

When an *obstacle* is in the path of flow (Fig 1.43), the flow can be studied by looking at two representative points 1 and 2. At point 1 the flow is *normal*, while at point 2 *stagnation* occurs and there is no flow. If we write the Bernoulli equation corresponding to the points 1 and 2, it has the form

$$P_1 + \frac{1}{2}\rho w_1^2 = P_2 + \frac{1}{2}\rho w_2^2$$

At point 2, $w_2 = 0$, and therefore

$$P_2 = P_1 + \frac{1}{2}\rho w_1^2$$

$$P_2 = P_{tot} = P_{stat} + P_{dyn}$$

P_2 is equal to the sum of the two pressures while $P_1 = P_{stat}$

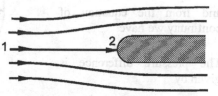

Fig 1.43 Fluid flow past an obstacle

4.3.2 Measurement of the static and total pressure
(Messung des statischen und des Gesamtdruckes)

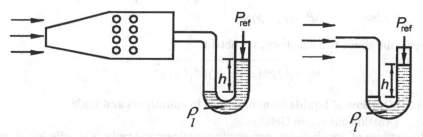

Fig 1.44(a) Static pressure measurement   (b) Total pressure measurement

The measurement of the static pressure P_{stat} may be carried out by using the device shown in Fig 1.44(a). This consists of a closed tube with *small holes* on the *side walls* connected to a *U-tube manometer*.

$$P_{stat} = P_{ref} + \rho_l gh$$

where ρ_l is the density of the liquid in the manometer.

The measurement of the total pressure can be carried out by using an *open tube* connected to a U-tube manometer as shown in Fig 1.44 (b). Here the *flow path is blocked* by the liquid in the U-tube and we have

$$P_{tot} = P_{ref} + \rho_l gh$$

The dynamic pressure P_{dyn} can be measured by *combining the two tubes* to give a difference measurement.

4.3.3 Venturi tube

(Venturirohr)

A venturi tube can be used to calculate the *flow velocity* in a flow tube by *measuring the pressures* at different points. Writing the Bernoulli eqn for the points 1 and 2, we have

$$P_1 + \frac{1}{2}\rho w_1^2 = P_2 + \frac{1}{2}\rho w_2^2$$

and from the equation of continuity we have

$$w_1 A_1 = w_2 A_2$$

The pressure difference is given by

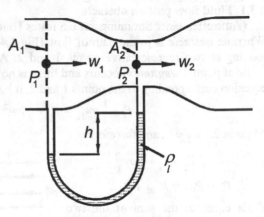

Fig 1.45 A venturi tube

$$\Delta P = P_1 - P_2 = \left(\frac{1}{2}\rho w_1^2\right)\left(\left(\frac{A_1}{A_2}\right)^2 - 1\right)$$

Also $\Delta P = (\rho_l - \rho)gh$

From the above two equations, we obtain

$$w_1 = \sqrt{2gh(\rho_l/\rho - 1)(A_1/A_2)^2}$$

4.3.4 Outflow of liquids from openings in containers and tanks

(Ausfluß aus einem Gefäß)

The outflow of liquids from openings in containers and tanks is usually less than the value that is *theoretically expected*, because the average velocity is less than the theoretical value of $w = \sqrt{2gh}$. The reduction in velocity is a consequence of the *internal friction* present in the liquid and the *friction of the walls* of the tank. The ratio of the actual velocity to the theoretical velocity is the *coefficient of velocity* φ.

Another reason for a reduction in the outflowing quantity of liquid is the **contraction of the flow stream** due to the sudden change in the direction of the flow. If the actual area of the opening is A, then the effective area of the opening is αA where α is the *coefficient of contraction*.

The product of the coefficient of velocity φ and the coefficient of contraction α is termed the *coefficient of discharge* μ.

$$\text{Coefficient of discharge} \quad \mu = \alpha\varphi$$

4.3.5 Open container with constant pressure head
(Offenes Gefäß mit konstanter Druckhöhe)

If the head of liquid is maintained constant at a value h, then we can assume

$$w_1 = 0 \quad \text{and} \quad P_1 = P_2 = P_0$$

where P_0 is the atmospheric pressure.

Using the Bernoulli equation we have

$$h_1 - h_2 = h = \frac{w_2^2}{2g}$$

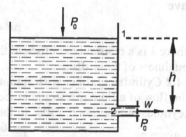

Fig 1.46 Open container with a constant pressure head

Using the continuity equation and taking into consideration the coefficient of discharge μ, we obtain the expression for the out flowing volume

$$q_V = \mu A \sqrt{2gh}$$

4.3.6 Container closed at the top with constant pressure head
(Geschlossenes Gefäß mit konstanter Druckhöhe)

In this case, the pressure at 1 is no longer the atmospheric pressure P_0, but some other value of pressure P_1. The pressure at the outflow opening is however P_0.

If we write the excess pressure $P_e = P_1 - P_0$, then we obtain using the Bernoulli equation an expression for the velocity of outflow

$$w = \varphi\sqrt{2\left(gh + \frac{P_e}{\rho}\right)}$$

and for the volume flow

$$q_V = \mu A \sqrt{2\left(gh + \frac{P_e}{\rho}\right)}$$

The quantity $\left(gh + \dfrac{P_e}{\rho}\right)$ can be found by using a manometer which is placed at the height of the output point.

4.4 Resistance to fluid flow in horizontal tubes
(Widerstände in Rohrleitungen)

4.4.1 Drop in pressure and the resistance coefficient
(Druckabfall und Widerstandszahl)

The flow of a fluid through a straight horizontal pipe is only possible when a *pressure gradient* exists along the length of the tube. This is *almost linear* and the pressure at the outlet is equal to the atmospheric pressure. In the case of *streamline* or *laminar flow* in a tube with *smooth walls*, the fall in pressure ΔP depends on the average velocity w. For a pipe of diameter d and length l we have

$$\Delta P = \lambda \frac{l\rho}{2d} w^2$$

where λ is a resistance coefficient that depends on the Reynolds number and the roughness of the walls.

4.4.2 Cylindrical tubes with smooth walls (Glattes Kreisrohr)

The flow of a fluid in a straight tube with *smooth walls remains laminar* up to a Reynolds number $Re = 2300$. In the laminar region the resistance coefficient λ is given by

$$\lambda = \frac{64}{Re} = \frac{\Delta P \, 2d}{w^2 \rho l}$$

Combining this with the relation $Re = \dfrac{w d \rho}{\eta}$

we have

$$\Delta P = 32 \frac{\eta \, w \, l}{d^2}$$

4.4.3 Cylindrical tubes with rough walls (Rauhes Kreisrohr)

For higher values of the Reynolds number, the resistance coefficient λ depends only on the *relative wall roughness* factor k/d. For granular roughness (in contrast to wavy roughness) the expression for λ is

$$\lambda = \frac{1}{\left[2 \lg\left(\dfrac{d}{k}\right) + 1.14\right]^2}$$

where d is the diameter in mm, and k a number corresponding to the roughness of the walls. The boundary region between laminar and turbulent flow presents serious difficulties but specific expressions are available in this region as well.

4.4.4 Valves and bends (Ventile und Krümmer)

In the case of valves and bends, the *loss in pressure* is dependent on the *average value* of the *flow velocity* and is given by

$$\Delta P = \zeta \frac{\rho}{2} w^2$$

The *resistance number* ζ depends on the *dimensions* involved.

II Strength of materials
(Festigkeitslehre)
1 Basic concepts (Grundkonzepte)
1.1 Scope of the strength of materials
(Bereich Festigkeitslehre)

The strength of materials is a subject which deals with the effect of *external forces* on *elastic bodies* (and structures) which are basically in a *state of equilibrium*. It can be considered to be a part of mechanics and is partly based on the *principles of statics*. In addition, it uses experimentally obtained data about the *changes* in the *dimensions* of elastic bodies when external forces act on them.

When a solid body is subjected to the action of external forces, it becomes *deformed*. If the forces are *relatively small*, the body regains its *original form* when the forces are removed. This type of deformation is called *elastic deformation*. If the forces exceed a critical value called the *elastic limit*, there is a *permanent change* in the dimensions of the body called *plastic deformation*. Further increases in the forces can result in the body becoming *fractured*.

The external forces create *stresses* in the body, which lead to dimensional changes which are called *strains*. The data which is available from previous experimental studies in the strength of materials can be used to *specify the dimensions* that a component needs to have, to remain within the elastic range and consequently *avoid* becoming *fractured*.

- The strength of materials is an *indispensable tool* in the hands of design engineers, who use the extensive information acquired about materials over the years, to *design structures* which are both *safe* and *economical*.

- A *safety margin* has to be allowed when a component is designed. It is necessary to ensure that factors like variations in the properties of materials used, defects in manufacture or slight overloading, will not lead to the component becoming damaged. A component can be *designed in many ways*. It is however *desirable* that it is designed in such a way that it can be produced at the *lowest possible cost*. The component has to be *safe to use* and *economical* to produce.

- *Tests* under *actual working conditions* are important once the component has been manufactured. The load that a component can be subjected to, is *fixed by the design*. However, it is desirable that the *correctness of the design* should be tested after manufacture. *Modifications* can then be carried out to *correct* any *deficiencies* that the component may have.

1.2 The method of sections and the free body (Schnittverfahren)
When *external forces* act on a body, *internal forces* are created inside the body in order that *equilibrium* may be maintained. These are a direct consequence of

the action of the external forces themselves. A convenient way of studying the internal forces in a body is by the ***method of sections***.

Consider the rod AB which is in equilibrium under the action of ***two equal*** and ***opposite*** forces F as shown in Fig 2.1(a). If we ***make a section*** at right angles to the axis through any point C in the rod, we obtain two parts AC and CB which can be called *free bodies*.

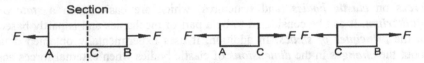

Fig 2.1(a) Rod before sectioning   (b) Two parts resulting from section at C

It is necessary to apply two forces F at C, in order to fulfil the requirement that both parts should be in equilibrium. From this we can see that ***internal forces*** of magnitude F are produced in rod AB, when external forces F act on it. In a free body the internal forces can be considered to have been ***replaced by external forces***. This facilitates the study of the problems involved in bodies which although in ***equilibrium***, are in a ***state of stress*** under the action of external forces.

1.3 Stress (Spannung)

1.3.1 Normal stress (Normalspannung)

The type of loading experienced by the rod in Fig 2.1 is called ***axial loading***, because the external forces acting on the rod are ***directed along the axis*** of the rod. Let the area of cross-section of the rod be A. The force per unit area of the cross-section is called the ***normal stress***, and is denoted by the greek letter σ.

$$\text{Normal stress } \sigma_{av} = \frac{F}{A}$$

The stress that exists under the conditions of axial (or longitudinal loading) is called the normal stress because it acts in a direction normal to the cross-section. Normal stress can be of two types (a) ***tensile*** or ***tensional stress*** and (b) ***compression*** or ***compressional stress*** (Fig 2.2).

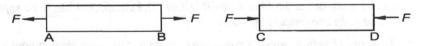

Fig 2.2 (a) Tensile stress   Fig 2.2 (b) Compressional stress

The stress taken over the whole cross-section is the ***average value*** σ_{av}. The stress may ***vary over the cross-section*** and to find the stress at any point we should consider the magnitude of the force dF acting over a small area dA surrounding the point. The internal force F acting over the cross-section is given by

$$F = \int dF = \int_A \sigma \, dA$$

1.3.2 Shear or shearing stress (Abscherspannung)

The type of stress which occurs when *transverse forces* are applied to a rod as shown in Fig 2.3 is called *shear or shearing stress.*

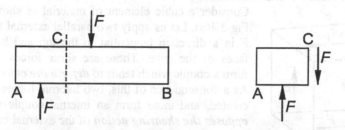

Fig 2.3(a) Transverse forces acting on a bar (b) Free body with section at C

If we make a section at a point C between the points of application of the two forces, we have the free body AC as shown in Fig 2.3(b). For equilibrium to be maintained, it is necessary for an internal force F to exist which is *tangential to* the *plane of the section*. Such a force is called a *shear (or a shearing)* force. Dividing the shear force by the area of cross-section A, we obtain the *average shear stress*. This is denoted by the greek letter τ and we can write

$$\tau_{av} = \frac{F}{A}$$

As stated in the case of the normal stress, the shear stress can also vary over the cross-section.

1.4 Strain (Verformung)

1.4.1 Longitudinal strain (Dehnung)

When a body is subjected to the action of external forces, it can change its form. The longitudinal strain which is denoted by the greek letter ε is equal to the ratio of the *increase in length* to the *original length* of the rod.

$$\text{Longitudinal strain } \varepsilon = \frac{\Delta l}{l}$$

1.4.2 Hooke´s Law (Hookesches Gesetz)

When a body is subjected to the action of external forces and the changes in form are *elastic*, then the *longitudinal strain* produced is *proportional* to the *longitudinal stress*. This relationship is called Hooke´s Law.

Stress σ = constant E × strain ε

$$\sigma = \frac{\Delta l}{l} E = \varepsilon E$$

E is a constant for a given material and is called the *modulus of elasticity* or the Young´s modulus.

The strain is a quantity which is **dimensionless** and therefore the modulus E has the same units as the stress. The stress and the Young's modulus are usually expressed in units of N/mm².

1.4.3 Shear strain (Schubverformung)

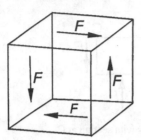

Fig 2.4 (a) Shear forces in a cube

Consider a cubic element of material as shown in Fig 2.4(a). Let us apply two parallel external forces F in a direction tangential to the top and bottom faces of the cube. These are shear forces which form a couple which tends to **deform the cube**.

As a consequence of this, two internal forces F are created, and these form an internal couple which **opposes the shearing action** of the external couple. The cube is in **equilibrium** under the action of the **external** and **internal couples**, and changes its shape to that of an **oblique parallelepiped,** the cross-section of which is shown in Fig 2.4 (b). The shear strain is equal to the angle γ.

For values of stress which are within the elastic limit, the shear strain γ is proportional to the stress.

We can therefore write for a **homogeneous isotropic** material

Shear stress $\tau = $ constant $G \times$ shear strain γ

or $\tau = G\gamma$

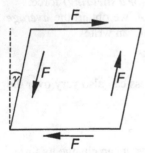

Fig 2.4 (b) Shear Strain

where G is a constant called the **modulus of rigidity** or the **shear modulus**. The angle γ is in radians and is dimensionless. The modulus G has therefore the same units as the stress τ, namely N/mm².

1.4.4 Poisson's ratio (Poisson's Zahl)

If a homogeneous slender bar is axially loaded, the Hooke's law is valid provided the elastic limit of the material is not exceeded. We can write using Hooke's law

$$\varepsilon_x = \sigma_x/E$$

where E is the modulus of elasticity (Young's modulus) of the material.
The **normal stresses** along the y and z-axes **are zero**.

$$\sigma_y = \sigma_z = 0.$$

However the **strains** ε_y and ε_z **are not zero**.

In all engineering materials, an **elongation produced in the axial** (or longitudinal) **direction,** is accompanied by a **contraction in the transverse** *direction* provided that the material is *homogeneous and isotropic*. In this case the strain is the *same for any transverse direction*.

Therefore $\varepsilon_y = \varepsilon_z$

This is called the lateral strain, and the *ratio* of the *lateral strain* to the *longitudinal strain* is called *Poisson's ratio* (indicated by the greek letter v).

v = Lateral strain / Longitudinal strain

or $v = -\dfrac{\varepsilon_y}{\varepsilon_x} = -\dfrac{\varepsilon_z}{\varepsilon_x}$

From this we can write expressions for the three strains as follows:

$$\varepsilon_x = \frac{\sigma_x}{E}, \quad \varepsilon_y = \varepsilon_z = -v\frac{\sigma_x}{E}$$

1.5 Types of Load (Beanspruchungsarten)

1.5.1 Tensional and compressional loads
(Zugbeanspruchung und Druckbeanspruchung)

These types of *axial load* have already been discussed in section 1.3. Both compressional and tensional loads are possible, and the stress which is called the normal stress acts in a direction *normal* to the *cross-section*.

1.5.2 Loads which cause bending (Biegebeanspruchung)

In pure bending, a bar is acted on by *two equal* and *opposite couples* acting in the same longitudinal plane. In many cases of bending however, *shear forces* also exist and need to be considered.

1.5.3 Loads which cause buckling (Knickbeanspruchung)

When an axial compressional load is applied to a bar, the bar can suddenly *buckle* when the load reaches a *certain critical value*. The shape of the deformed bar may appear to be similar in the cases of bending and buckling, but there is clearly a *basic difference* between the two cases. *Bending* is caused by *stress* in the material, while *buckling* is caused by *instability*.

1.5.4 Shear loads (Abscherbeanspruchung)

In the case of shear loads, the external forces act in a direction which is *transverse to the axis* and tend to push successive layers of material in a transverse direction. *Internal transverse forces* are created which tend to resist the change in shape. These internal forces are called *shear forces*.

1.5.5 Torsional loads (Verdrehbeanspruchung)

The external forces corresponding to torsional loads, are in the form of an external *couple* which *twists* a rod about its axis. This results in an internal couple being created which *opposes the twisting moment* of the external couple.

1.6 Strength (Festigkeit)

1.6.1 Concept of strength (Begriff der Festigkeit)

The meaning of the word strength in the context of the strength of materials, refers to the ability of a material or a component to *resist fracture* under given conditions of mechanical loading.

Tests to determine the strength of a material are carried out under standard conditions. A *numerical value* is obtained, which gives a good *indication of the strength* of a material and enables us to compare strengths of different materials.

1.6.2 Determination of strength under conditions of static loading
(Festigkeit bei statischer Belastung)

The specimen to be tested is carefully made to *standard dimensions*. It is then placed in a test machine and subjected to a *gradually increasing tensional load*, starting at zero and ending at a value for which the specimen breaks. The values of *stress and strain* are found for various loads, and a *graph* between stress and strain is plotted as shown in Fig 2.5(a) for annealed soft steel.

In the region OP, the strain is *proportional* to the stress and Hooke's law holds. Just above this is the point E, which corresponds to the *elastic limit*. If the load is removed at this stage, the specimen reverts to its *original shape* and *size*. The behaviour of the material so far has been elastic, and the material is said to have undergone *elastic deformation*.

If the load is further increased, a stage is reached at Y, where a *sudden increase* in *length* takes place *without any increase in load*. The stress corresponding to Y is called the *yield strength*. Beyond the point Y, the deformation is *no longer proportional* and *elastic*. This deformation which is called a *plastic deformation* is permanent. In the region YB, the specimen *elongates permanently* and its cross-section becomes *reduced uniformly* along its length.

The material becomes *stronger* and is said to be *strain hardened* or *work hardened*. Beyond point B which corresponds to the *maximum possible load*, the cross-section *decreases* only at its *weakest point*. This is called *necking*. The stress at B is called the *ultimate tensile strength* (UTS). The specimen finally *breaks (or fractures)* at C and this stress is called the *breaking strength*.

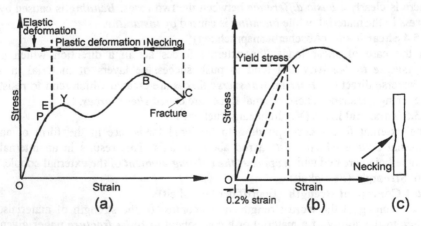

Fig 2.5 (a) Stress vs strain curve for   (b) Stress vs strain curve for   (c) necking
annealed steel   hardened steel

A curve of the type shown in Fig 3.3 (a) which shows a *clear change* at the *yield stress* is only observed in *some materials* like soft steel. In the case of materials like hardened steel, the curve is as shown in Fig 3.3 (b).This curve is a smooth one, which has neither a *definite elastic limit nor a yield stress*. In such cases, a different method has to be adopted to find the yield strength. The method that has been *universally adopted*, has been to specify the stress corresponding to a *permanent extension* of 0.2% as the yield strength. This can be found by drawing a straight line through the point corresponding to 0.2% strain. The straight line is drawn parallel to the initial part of the curve which is the proportional region. This meets the curve at the point Y which corresponds to the yield stress. The modulus of elasticity (or the Young's modulus) can be found from the slope of the curve in the proportional region.

> **Young's modulus E = Stress/Strain (N/mm^2)**

1.6.3 Strength under dynamic loading conditions
(Festigkeit bei dynamischer Belastung)
The test values obtained for the strength of materials obtained under conditions of *static loading* may not be useable in many types of applications. A good example is the case of materials used in the construction of machines which are subjected to *repeated stress*.
In these cases it is necessary to carry out a dynamic test where a polished sample of the material is subjected to a *dynamically varying load*. This gives a value for the *fatigue strength* which is usually less than the static value. It may also be both *desirable* and *necessary* to test not only a standard specimen of the material, but also the *actual component* under dynamic loading conditions.

1.6.4 Effects of irregularities in the cross-section (Kerbwirkung)
The cross-section of a component is rarely uniform like the cross-section of a test sample. Components often have notches, channels, splines and other irregularities which cause the *stress to be nonuniform* across the cross-section. This means that the component has to withstand higher stresses and has therefore to be made of stronger material or of larger cross-section.

1.7 Allowable stress and the safety factor
(Zulässige Spannung und Sicherheitsfaktor)
When a component is designed, the load that it is allowed to carry under operating conditions is called the *allowable load*. For reasons of safety, this allowable load must be well below the *ultimate load* which is the load under which the *component will break*.
A *safety factor* v has been defined in terms of the ratio given below, but can also be defined in terms of other similar ratios.

$$\text{Safety factor } v = \frac{\text{ultimate stress}}{\text{allowable stress}}$$

The determination of a safety factor which is useful and relevant under a certain set of conditions is a very important engineering task. In addition to the above considerations, material variations, fatigue and the effect of other components in the structure must be considered.

If the safety factor is *too small*, the *chances of failure* may become *unacceptably large*. If the safety factor is *too large*, the component would become too *uneconomical* to make. High safety factors may not be acceptable in aircraft design where weight is an important factor.

The *safety factor for steel* is of the order of 1.5. Materials like grey cast iron are extremely brittle. Here the value for the breaking stress can be used instead of the value for the ultimate stress. A safety factor of 2 is appropriate for cast iron.

2 Bending loads (Biegebeanspruchung)
2.1 Pure bending (Reine Biegung)
2.1.1 Beams of uniform cross-section
(Träger mit gleich bleibendem Querschnitt)

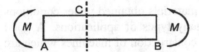 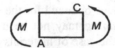

Fig 2.6 (a) Pure bending of a uniform beam   (b) Free body after sectioning at C

Consider a beam of uniform cross-section being subjected to the action of two *equal and opposite couples* M as shown in Fig 2.6 (a). The couples act in the same longitudinal plane and the beam is symmetric with respect to the plane of the couples. When a beam bends under these conditions, it is said to undergo *pure bending*.

If we make a section in the beam at C, it is clear that the forces exerted on AC by the portion CB must be equivalent to a couple as shown in Fig 2.6(b). Thus the internal forces acting at any cross-section when pure bending occurs are equivalent to a couple. The moment M of the couple is known as the *bending moment* at this cross-section. Since the point C chosen by us is an arbitrary point, it follows that when a beam undergoes *pure bending*, the *bending moment* is the *same at all cross-sections* and has the value M.

Example
An example of a beam being subjected to pure bending is shown in Fig 2.7. Making a section at any arbitrary point E, we can verify that the internal forces acting at any cross-section at any point E between D and C are equivalent to a couple of moment 40kNm.

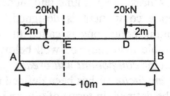

Fig 2.7 Pure bending of a transversely loaded beam

2.1.2 Deformations produced in a beam undergoing pure bending
(Verformungen in einem Träger)

Consider a uniform beam having a plane of symmetry which is subjected at its ends to equal and opposite couples M acting in the plane of symmetry (Fig 2.8). The beam will bend under the action of the couples, but will remain symmetric with respect to the longitudinal plane in which the couples act. Since the *bending moment* is the *same* at any cross-section, the *beam will bend uniformly* (Fig 2.8). Thus the lines AB and A'B'(not shown) which were *originally straight lines*, will now become *arcs of circles* with centre C.

We can see that while AB has decreased in length, the length of A'B' has increased (not shown).Therefore there must be a surface which is parallel to the upper and lower surfaces where the change in length is zero. This is called the *neutral surface*. The neutral surface intersects the plane of symmetry along an arc of a circle, and intersects the transverse section along a straight line called the *neutral axis*. Two sections, a *longitudinal section* (the plane of symmetry) and a *transverse section* are shown in

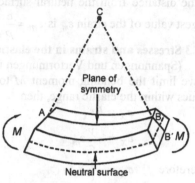

Fig 2.8 Beam undergoing pure bending

Fig 2.9. The *neutral surface* intersects the *plane of symmetry* along the *arc of the circle* DE as shown in Fig 2.9(a) and the *transverse section* along a *straight line* called the *neutral axis* (Fig 2.9 (b)).

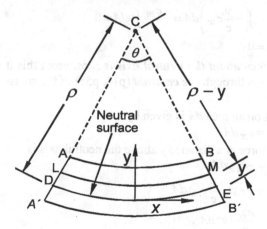

Fig 2.9 (a) Longitudinal vertical section
(plane of symmetry)

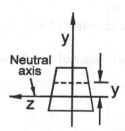

Fig 2.9 (b) Transverse section

If the radius of the arc DE is ρ and the angle subtended at the centre θ, then
$$l = \rho\theta$$
where l is the length of DE. This is also the **original length** of the **undeformed bar**. If we consider an arc LM at a distance y above the neutral surface and denote its length as l', then $\quad l' = (\rho - y)\theta$

the change in length δ is $\qquad \delta = l' - l = (\rho - y)\theta - \rho\theta \quad$ or $\quad \delta = -y\theta$

The longitudinal strain is $\qquad \varepsilon_x = \dfrac{\delta}{l} = -\dfrac{y\theta}{\rho\theta} \quad$ and therefore $\quad \varepsilon_x = -\dfrac{y}{\rho}$

If the distance from the neutral surface to the top of the beam is c, then the largest value of the strain ε_m is $\varepsilon_m = \dfrac{c}{\rho}$

2.1.3 Stresses and strains in the elastic range
(Spannungen und Verformungen im elastischen Bereich)
If we limit the bending moment M to values within the elastic range, then

$$\sigma_x = E\varepsilon_x$$

$$\varepsilon_x = -\frac{y}{c}\varepsilon_m$$

Therefore $\qquad \sigma_x = -\dfrac{y}{c}\sigma_m$

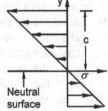

The normal stress is proportional to the distance from the neutral surface as shown in Fig 2.10. In the case of pure

Fig 2.10 Distribution of normal stress in a bent beam

bending there can be no net force in the longitudinal direction which means that
$$\int \sigma_x dA = 0 \quad \text{for any cross-section.}$$

$$\sigma_x\, dA = \int\left(-\frac{y}{c}\sigma_m\right)dA = -\frac{\sigma_m}{c}\int y\, dA = 0$$

which gives $\qquad \int y\, dA = 0$

The **first moment of the area** taken **about the neutral axis is zero**. From this it follows that the **neutral axis passes** through **the centroid** (p14, p58) of the cross-section.

The elementary force dF_x acting on an area dA is given by
$$dF_x = \sigma_x\, dA$$
The moment of this elementary force at a distance y above the neutral axis is
$$dM = -\sigma_x\, y\, dA$$

$$M = \int -y\left(-\frac{y}{c}\sigma_m\right)dA$$

$$M = \frac{\sigma_m}{c}\int y^2\, dA \ .$$

If we write $I = \int y^2\, dA$

The integral I is the **second moment of an area** (p58) for the cross-section with respect to the neutral axis.

Therefore
$$\sigma_m = \frac{Mc}{I}$$

The ratio $\dfrac{I}{c}$ is dependent only on the **geometry of the cross-section**. This ratio is termed the **elastic section modulus S**.

Elastic section modulus $S = \dfrac{I}{c}$

Therefore $\sigma_m = \dfrac{M}{S}$

The maximum stress σ_m can be seen to be **inversely proportional to the section modulus S**. From this it is clear that it is an advantage to design beams with **as large a value** of S as is practical so that the **maximum stress is kept low**. For a beam with a rectangular cross-section having a width b and a depth h.

$$S = \frac{I}{c} = \frac{bh^3/12}{h/2} = \frac{1}{6}bh^2 = \frac{1}{6}Ah \qquad (A = \text{area of cross-section})$$

If two beams have the same area of cross-section as in Fig 2.11 (a) and (b), the **beam of larger depth** (b) is **able to resist bending better** than (a). Also **wide flanged beams** with cross-section as shown in Fig 2.11(c) are able to **resist bending better** than other shapes. For a given area of cross-section and a given depth, these beams have a large part of their cross-section located away

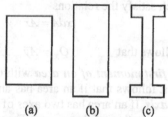

(a)   (b)   (c)
Fig 2.11 Beams with different cross-sections

from the neutral axis. This shape provides large values of I and S.

2.1.4 Radius of curvature and curvature (Krümmungsradius und Krümmung)
As seen in section 2.1.2 when a uniform beam is subjected to pure bending, the beam bends to form the arc of a circle. The radius of curvature ρ was given by

$$\varepsilon_m = \frac{c}{\rho}$$

The term **curvature** is used for the **reciprocal** of the radius of curvature ρ.

$$\text{Curvature} = \frac{1}{\rho} = \frac{\varepsilon_m}{c}$$

In the elastic range $\varepsilon_m = \dfrac{\sigma_m}{E}$

Substituting we have $\dfrac{1}{\rho} = \dfrac{\sigma_m}{Ec} = \dfrac{1}{Ec}\dfrac{Mc}{I}$

$$\text{Curvature} = \frac{1}{\rho} = \frac{M}{EI}$$

2.2 First and second moments of an area
(Flächenmoment 1.Grades und Flächenmoment 2.Grades)

2.2.1 First moment of an area and the centroid
(Flächenmoment 1.Grades und Flächenschwerpunkt)

Consider an area A which is in the xy plane. If the coordinates of a small element of area dA are x and y, then the first moment of the area A with respect to the x axis is defined as the integral

$$Q_x = \int_A y\,dA$$

Similarly, the first moment of area can be defined with respect to the y axis as the integral

$$Q_y = \int_A x\,dA$$

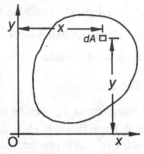

Fig 2.12 First moment of an area

The **centroid** of the area is defined as the point C having coordinates $\bar{x}$ and $\bar{y}$ which satisfy the relations

$$\int_A x\,dA = A\bar{x} \quad \text{and} \quad \int_A y\,dA = A\bar{y}$$

It follows that $\quad\quad Q_x = A\bar{y} \quad$ and $\quad\quad Q_y = A\bar{x}$

The **first moment of an area** with respect to an **axis of symmetry** is **zero**. From this it follows that if an area has an **axis of symmetry**, the **centroid must be on this axis**. If an area has two axes of symmetry, the **centroid** must lie on the **point of intersection** of these axes. Therefore the centroids of a rectangle or a circle must coincide with their **geometric centres**.

2.2.2 Second moment of an area (or areal moment of inertia)
(Flächenmoment 2. Grades)

The moment of inertia of a solid body is defined by the integral $J = \int y^2 dm$.

A similar quantity can be defined with respect to an area and is called the **second moment of an area** (or the areal moment of inertia). If we have an area A in the xy plane as shown in Fig 2.13, the second moments of the area with respect to the x and y axes are defined by the integrals

$$I_x = \int_A y^2 dA \quad \text{and} \quad I_y = \int_A x^2 dA$$

These integrals are called rectangular moments of inertia.

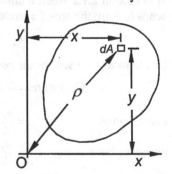

Fig 2.13 Second moment of an area

The *polar second moment of an area* A with respect to point O is defined by

$$I_0 = \int_A \rho^2 \, dA = \int_A (x^2 + y^2) \, dA$$

It follows that

$$I_0 = I_x + I_y$$

where ρ is the distance of dA from the origin O.

The different *radii of gyration* for an area are defined by the following relations.

$$I_x = r_x^2 \, A, \quad I_y = r_y^2 \, A, \quad I_0 = r_0^2 \, A$$

Therefore

$$r_0^2 = r_x^2 + r_y^2$$

2.2.3 Parallel axis theorem (Steinersche Verschiebesatz)

The parallel axis theorem for moments
of inertia (proved on p22) also holds
for second moments of an area and will
only be stated here.
Let I_x be the second moment of an
area A with respect to an arbitrary x
axis and $I_{x'}$ the second moment of the
same area with respect to a parallel
x' axis passing through the centroid C.
If the separation between the two axes
is l, then we can write

$$I_x = I_{x'} + Al^2$$

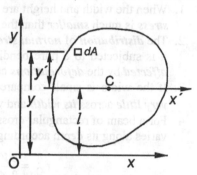

Fig 2.14 Parallel axis theorem

A similar formula holds for the *polar moment of inertia* as given below.

$$I_0 = I_c + Al^2$$

Here I_0 and I_c are the second moments of the area A with respect to an
arbitrary point O and with respect to its centroid C. The distance OC $= l$.

2.3 Transverse loading and shearing stresses

(Querkraftbiegung und Schubspannung)

When uniform beams are in a horizontal
position and have vertical loads, then the
loading is transverse.
A *cantilever* beam is shown in Fig 2.15.
This has one fixed end B. A single force
F acts at the free end A. Let us make a
section at C at a distance x from A and
consider the free-body diagram of the
part AC. The internal forces acting on
AC must be equivalent to a shearing

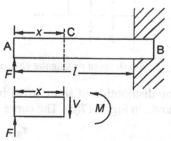

Fig 2.15 Stresses in a cantilever

force V of magnitude $V = F$ and a couple
M of bending moment $M = Fx$.

As seen previously, the stress normal to
the cross-section is required to form the
bending moment. Shearing stresses are
also present and we can write for the

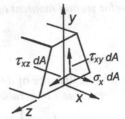

    y direction $\int \tau_{xy}\, dA = -V$ and

for

the z direction $\int \tau_{xz}\, dA = 0$

Fig 2.16 Stress components in a
beam

An analysis of the shearing stresses in a beam are beyond the scope of this book.
However a summary of some of the the most important features is given below.

1. When the width and height are small compared to the length, the ***shearing
   stress*** is much ***smaller*** than the ***normal stress***, typically less than 10%.
2. The ***distribution*** of ***normal stresses*** in a cross-section is the same as when
   it is subjected to a pure bending couple of moment $M = Fx$ and ***is not
   affected*** by the ***deformations*** caused by the ***shearing stresses***.
3. If the width is small compared to the depth, the ***shearing stress varies
   very little*** across ***its width*** and we may replace it by the average stress τ_{av}.
4. For a beam of rectangular cross-section as shown in Fig 2.18(a), the stress
   varies along its depth according to the formula given below.

$$\tau_{xy} = \frac{3}{2}\frac{V}{A}\left(1 - \frac{y^2}{c^2}\right)$$

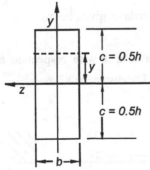

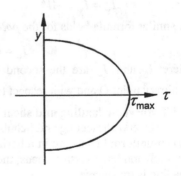

Fig 2.17(a) Beam of rectangular cross-section   Fig 2.17 (b) Distribution of shear stress

The distribution of shear stress in a transverse section of a rectangular beam is
shown in Fig 2.17(b). The curve is parabolic and the maximum stress is

$$\tau_{max} = \frac{3}{2}\frac{V}{A}$$

2.4 The deflection of beams (Durchbiegung von Trägern)
2.4.1 Determination of the deflection by integration
(Integrationsmethode für die Bestimmung von Durchbiegung)

It has been seen that if a beam of narrow cross-section is subjected to pure bending, it is bent into an arc of a circle. The curvature of the neutral surface (when the bending is within the elastic range) is given by the expression

$$\text{Curvature} \quad \frac{1}{\rho} = \frac{M}{EI}$$

The curvature at a point P (x, y) on a plane curve is given by the expression

$$\text{Curvature} \quad \frac{1}{\rho} = \frac{\dfrac{d^2 y}{dx^2}}{\left[1 + \left(\dfrac{dy}{dx}\right)^2\right]^{3/2}}$$

In the case of bending in beams, the gradient dy/dx is very small and the square of this is much smaller than unity and can be neglected. Therefore

$$\text{Curvature} \quad \frac{1}{\rho} = \frac{d^2 y}{dx^2}$$

EI is known as the ***flexural rigidity*** and is ***constant*** for ***a uniform beam***.

2.4.2 Example: Deflection of a cantilever by the method of integration
(Durchbiegung eines Freiträgers mit Einzellast)

Consider the cantilever beam AB of length l having a load F at its free end. The bending moment and the curvature of the beam vary from cross-section to cross-section. For a section at a distance x from A we can write

$$\frac{1}{\rho} = \frac{d^2 y}{dx^2} = -\frac{Fx}{EI}$$

The curvature of the neutral surface varies linearly with x from zero at A to $-Fl/EI$ at B.

$$EI \frac{d^2 y}{dx^2} = -Fx$$

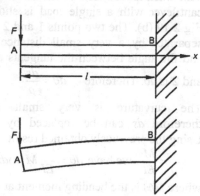

Fig 2.18 A cantilever

Integrating with respect to x, we have

$$EI \frac{dy}{dx} = -\frac{1}{2} Fx^2 + c_1$$

To find c_1 we use the boundary condition at the fixed end, where $x = l, \dfrac{dy}{dx} = 0$. Substituting, we obtain $c_1 = \dfrac{1}{2} Fl^2$

Therefore $\qquad\qquad EI\dfrac{dy}{dx}=-\dfrac{1}{2}Fx^2+\dfrac{1}{2}Fl^2$

Integrating both sides of the equation with respect to x, we have

$$EI\,y=-\dfrac{1}{6}Fx^3+\dfrac{1}{2}Fl^2x+c_2$$

Using the boundary condition for the fixed end which is $x=l, y=0$, we have

$$0=-\dfrac{1}{6}Fl^3+\dfrac{1}{2}Fl^3+c_2$$

Therefore $\qquad\qquad c_2=-\dfrac{1}{3}Fl^3$

Substituting for c_2 we have $\quad EI\,y=-\dfrac{1}{6}Fx^3+\dfrac{1}{2}Fl^2x-\dfrac{1}{3}Fl^3$

or $\qquad\qquad y=\dfrac{F}{6EI}\left(-x^3+3l^2x-2l^3\right)$

The deflection and slope at the free end can be found by substituting $x=0$.

From this we find $\quad y_A=-\dfrac{Fl^3}{3EI}\qquad$ and $\qquad \theta_A=\left(\dfrac{dy}{dx}\right)_A=\dfrac{Fl^2}{2EI}$

2.4.3 The moment-area method for finding the deflection of beams
(Momentenflächen Methode für die Bestimmung von Durchbiegung)

This method uses the *geometric properties of the elastic curve* (which is the curve into which the axis of the beam is bent when loaded). The elastic curve of a cantilever with a single load is shown Fig 2.19 (b). The two points 1 and 2 are separated by a very small distance ds and the angle between the tangents at 1

and 2 is $d\theta$. Therefore $\quad d\theta=\dfrac{ds}{\rho}$

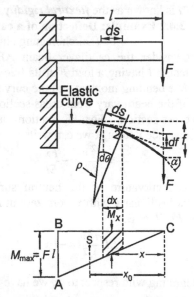

The curvature is very small and therefore ds can be replaced by dx. Using the previously obtained result

$\dfrac{1}{\rho}=\dfrac{M}{EI}$, we have $d\theta=\dfrac{1}{EI}M(x)dx$

where $M(x)$ is the bending moment at x. If we plot $M(x)$ vs x as in Fig 2.19 (c), $M(x)dx$ is the shaded area and therefore

$dA_M=M(x)\,dx\quad$ and $\quad d\theta=\dfrac{1}{EI}dA_M$

where A_M = total area under the curve.

Fig 2.19 Cantilever with elastic curve and moment-area diagram

Deflection angle $\alpha = \int d\theta = \dfrac{1}{EI}\int dA_M$

$$\alpha = \frac{A_M}{EI} = \frac{1}{EI}\frac{1}{2}M_{max} = \frac{1}{2}\frac{Fl^2}{EI}$$

Vertical deflection $df = x d\theta = \dfrac{M_x}{EI} x dx$

The element of area $dA_M = M_x dx$ is shown shaded in Fig 2.19 (b) and $M_x x dx$ is the moment of this about a vertical axis at the load end. The sum of all the elementary moments is $\int M_x x dx = A_M x_0$ where A_M is the total area under the curve and x_0 the distance from the vertical axis to the centroid S of the area A_M.

Total vertical deflection $f = \dfrac{1}{EI} A_M x_0$

For a cantilever $A_M = Fl\dfrac{l}{2} = \dfrac{Fl^2}{2}$ and $x_0 = \dfrac{2}{3}l$

$$f = \frac{1}{EI}\frac{Fl^2}{2}\frac{2}{3}l = \frac{Fl^3}{3EI}\quad\text{and}\quad \tan\alpha = \frac{A_M}{EI} = \frac{Fl^2}{2EI}$$

which are the same as the values obtained by integration in section 2.4.2

---

3 The buckling of columns (Knickung)

When a bar or column is *axially loaded* and the load is gradually increased, failure occurs in relatively short columns when the *compressional stress exceeds* a *certain value* for a given material. However, when the length of the column is large in comparison with its transverse dimensions, failure occurs by *buckling* (or transverse bending) when the stress reaches a *critical value* σ_{cr} which is *well below* the *allowable compressional stress* σ_{all}. i.e., $\sigma_{cr} < \sigma_{all}$.

3.1 Pin-ended columns (Beidseitig gelenkig gelagerter Druckstab)

We wish to find the critical value σ_{cr} for which a column ceases to be stable. If the applied stress $\sigma > \sigma_{cr}$, then the *slightest misalignment* or *disturbance* causes the column to *buckle*. A column AB of length l which supports an axial load F and is pin-connected at both ends is shown in Fig 2.20. Consider a point Q on the elastic curve of the beam at a distance x from the end A. The lateral deflection of Q is denoted by y and the bending moment at Q is $M = -Fy$.

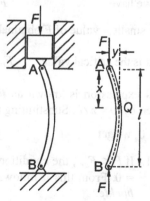

Fig 2.20 A column with pins on both ends

Substituting in the equation derived in section 2.4.1

$$\frac{d^2y}{dx^2} = \frac{M}{EI}$$

we have

$$\frac{d^2y}{dx^2} = -\frac{F}{EI}y$$

or

$$\frac{d^2y}{dx^2} + \frac{F}{EI}y = 0$$

This is a linear *homogeneous differential equation* with constant coefficients. If we write $p^2 = \dfrac{F}{EI}$, then

$$\frac{d^2y}{dx^2} + p^2y = 0$$

The general solution of this differential equation is given by

$$y = A\sin px + B\cos px$$

We now apply the *boundary conditions* which have to be satisfied.

At end A of the column $x = 0$ and $y = 0$ which gives $B = 0$.

At end B of the column $x = l$ and $y = 0$ which gives

$$A\sin pl = 0$$

This condition is satisfied if $A = 0$ or $\sin pl = 0$.

1. If $A = 0$, then $y = 0$. *The column remains straight* and *does not buckle.*

2. If $\sin pl = 0$, $pl = n\pi$. Substituting for p in the expression $p^2 = \dfrac{F}{EI}$,

    we have

$$F = \frac{n^2\pi^2 EI}{l^2}$$

The smallest value of F obtainable from this expression corresponds to $n = 1$.

This is the critical value $F_{cr} = \dfrac{\pi^2 EI}{l^2}$

This expression is known as *Euler's formula*. The value of p for this case is given by $p = \pi/l$. Substituting this value in the equation for y and assuming that

$B = 0$, we get $y = A\sin\dfrac{\pi x}{l}$

1. If $F < F_{cr}$, the condition $\sin pl = 0$ is not satisfied and we assume that $A = 0$. From this it follows that the *column remains straight* and *does not buckle.*

2. If $F > F_{cr}$, the column *buckles* (or undergoes *a lateral deflection*) and the elastic curve has the form of a *sine curve.*

The magnitude of the stress that corresponds to the *critical load* is called the *critical stress* σ_{cr}. If we write $I = Ar^2$ where r is the radius of gyration and A the area of cross-section, we have

$$\sigma_{cr} = \frac{F_{cr}}{A} = \frac{\pi^2 EAr^2}{Al^2} = \frac{\pi^2 E}{(l/r)^2}$$

The quantity l/r is termed the *slenderness ratio* of the column. A graph of σ_{cr} vs l/r is shown in Fig 2.21 for steel. It is assumed that $E = 200$ GPa and (the stress corresponding to) the yield strength$\sigma_Y = 250$ MPa. The region for which $\sigma_{cr} > \sigma_Y$ need not be considered because this is *outside the elastic range*. The Euler formula leads us to the conclusion that the critical load at which buckling occurs *does not depend* on the *strength of the material*. It depends only on the *modulus of elasticity* and the *slenderness ratio*.

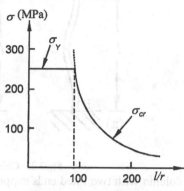

Fig 2.21 Graph of σ_{cr} vs l/r

Two slender columns having the same dimensions one of *high tensile strength steel* and the other of *mild steel* will buckle under the *same conditions of load*. It can also be seen that F_{cr} can be increased by increasing I. This can be achieved without *changing the cross-section* by using *tubular* rather than solid columns.

3.2 Column with one fixed end and one free end

(Einseitig eingespannter Druckstab) Consider a column with one fixed end B and one free end A which supports a load F (Fig 2.22). The column behaves like the *upper half of a pin-ended column*. The critical load in this case is the same as for a pin-ended column with length equal to twice the length l of the given column. The effective length $l_e = 2l$ and substituting in Euler's

formula $\quad F_{cr} = \dfrac{\pi^2 EI}{l_e^2}$

and $\quad \sigma_{cr} = \dfrac{\pi^2 E}{(l_e/r)^2}$

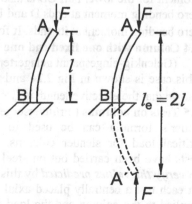

Fig 2.22 Column with one fixed end and one free end

3.3 Column with two fixed ends (Beidseitig eingespannter Druckstab)

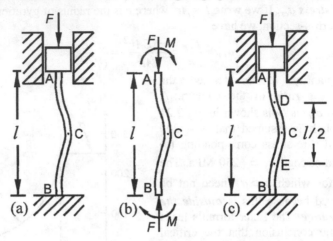

Fig 2.23 Column with two fixed ends

A column with two fixed ends supporting a load F is shown in Fig 2.23 (a). The symmetry of the supports and the conditions of the load requires that the *shear force* at C and *the horizontal components* of the *reactions* at A and B *should be zero* (Fig 2.23(b)). If we consider the upper half AC, the forces exerted on AC by the support at A and by the lower half CB must be identical. From this it follows that AC is *symmetrical about its midpoint* D (Fig 2.23(c)). This is a *point of inflection* where the *bending moment is zero.* Similarly the bending moment for the lower half CB is also zero at its midpoint E. The portion DE has zero bending moment at ends D and E and is similar to a pin-ended column with zero bending moment at its ends. It follows that the *effective length* is $l_e = l/2$.

3.4 Column with one fixed end one pin connected end

(Gelenkig-eingespannt gelagerter Druckstab)
This case is shown in Fig 2.24 and it can be shown that the effective length is $l_e = 0.7l$.

3.5 Tests on columns (Prüfungsergebnisse)

Euler's formula can be used to find the critical load for slender columns. Practical tests have been carried out on steel columns to *verify the values predicted* by this formula. In each test a centrally placed axial load was applied to the column and the load increased until failure occurred. The results of a large number of tests are shown in Fig 2.25. The results show a lot of scatter, but definite

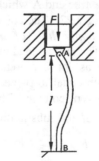

Fig 2.24 Column with one fixed and one pin connected end.

conclusions can be drawn.

1. In the case of *long columns*, Euler's formula *closely predicts the correct value* of σ_{cr} at which failure occurs corresponding to a given value of slenderness ratio l_e / r.

   The value of σ_{cr} *depends on the modulus of elasticity E, but not on the yield stress* σ_Y.

2. Failure occurs *in short columns* at the *yield stress* and we can write

   $$\sigma_{cr} \approx \sigma_Y$$

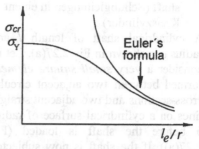

Fig 2.25 Results of load tests

3. In columns of *intermediate length*, failure occurs at *values of stress* which depend on both E and σ_Y.

Empirical formulae which give the values of allowable stress or critical stress corresponding to a given slenderness ratio have been used for a long time and have been refined with the passage of time. As may be expected, a single formula is not adequate for all applications. *Several formulae* each applicable to a range of values of slenderness ratio have been developed for different metals.

4 Torsion (Verdrehung)

Shafts of circular cross-section are used in many engineering applications. When such shafts are subjected to *twisting couples* or *torques*, they are said to be in *torsion*.

4.1 Torsion in a cylindrical shaft (Torsion in einem Kreiszylinder)

Consider a cylindrical shaft which is *fixed* at *one end* B as shown in Fig 2.26 (a). If a torque T is applied to the other end A (called the *free end*), this end will rotate through an angle φ which is called the *angle of twist* (Fig 2.26(b)). Experiment shows that if we operate within the elastic region

1. The angle of twist φ is proportional to the torque T.
2. The angle is also proportional to the length l of the shaft.

It is further observed that the *shape* of the shaft does *not change* when it is in a state of torsion. Different cross-sections of the shaft rotate through different angles, but each retains its *original shape*.

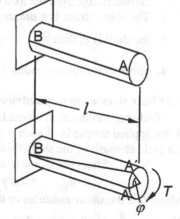

Fig 2.26 (a)&(b) Torsion in a cylindrical shaft

4.2. Shear strains in a cylindrical shaft (Schubgleitungen in einem Kreiszylinder)

A cylindrical shaft of length l and radius c is shown in Fig 2.27(a). Let us consider a very **small square element** formed between two adjacent circular cross-sections and two adjacent straight lines on a cylindrical surface of radius ρ before the shaft is loaded (Fig 2.27(a)). If the shaft is now subjected to a torsional load, the square is **deformed into a rhombus** (Fig 2.27(b)). It has been stated in section 1.4.3 that the **shear strain** γ in a given element is equal to the **change in the angles** which the sides of the element undergo when subjected to a shear stress.

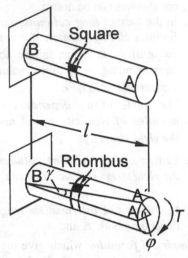

Fig 2.27 Shear strain in a cylindrical shaft

The angle γ in this case is the angle between the lines AB and A′B. Since the angle is small, the length $AA' = l\gamma$. But AA′is also equal to $\rho\varphi$

and therefore $\qquad \gamma = \dfrac{\rho\varphi}{l} \qquad$ where γ and φ are in radians.

This shows that

1. The shear strain γ is **proportional to the angle of twist** φ.
2. The shear strain γ is also **proportional to the distance** ρ of the point under consideration from the axis of the shaft.
3. The shear strain is a **maximum on the outer surface**, and if the radius of the shaft is c, then $\qquad \gamma_{max} = \dfrac{c\varphi}{l}$
4. Shear strain for any radius ρ is $\gamma = \dfrac{\rho}{c}\gamma_{max}$

4.3 Shear stresses in a cylindrical shaft

(Schubspannungen in einem Kreiszylinder)

If the applied torque is of such a magnitude that the shear stresses remain below the yield strength of the shaft, then Hooke′s law applies and we can write

Shear stress $\tau = G \times$ shear strain γ

or $\qquad \tau = G\gamma$

where G is the **shear modulus** or the **modulus of rigidity** of the material.

Since $\gamma = \dfrac{\rho}{c}\gamma_{max}$ it follows that $\tau = \dfrac{\rho}{c}\tau_{max}$ and therefore τ varies linearly with ρ.

The sum of the moments of the elementary shearing forces acting on any cross-section of the shaft must be equal to the torque T exerted on the shaft.

$$dF = \tau dA \qquad \text{and} \qquad T = \int \rho (\tau dA)$$

Substituting $\tau = \dfrac{\rho}{c}\tau_{max}$ in the above equation, we have $T = \dfrac{\tau_{max}}{c}\int \rho^2 dA$

$\int \rho^2 dA$ is the *polar moment of inertia* I_0 of the cross-section with respect to its

centre O. Therefore $\qquad T = \dfrac{\tau_{max}I_0}{c}$ or $\qquad \boxed{\tau_{max} = \dfrac{T}{I_0}c}$

Substituting for τ_{max}, we express the shear stress at any distance ρ from the axis

of the shaft as $\qquad \boxed{\tau = \dfrac{T}{I_0}\rho}$

The above two equations in boxes are known as the *elastic torsion formulae*.

4.4 Shafts for the transmission of power (Stäbe für Leistungsübertragung)
The polar moments of inertia for the area of cross-section of solid and hollow cylindrical shafts are given by

$$I_0 = \frac{1}{2}\pi c^4 \text{ for solid shafts and } I_0 = \frac{1}{2}\pi\left(c_2^4 - c_1^4\right) \text{ for hollow shafts.}$$

The *material* and the *cross-section* has to be selected so that the *maximum shearing stress* allowed *will not be exceeded.* The specific requirements to be satisfied are the *power transmitted* and the *speed of rotation.* The power associated with a body which is rotating under the action of a torque T is

$$P = T\omega$$

Since $\omega = 2\pi f$, we can write $P = 2\pi f T$ and $T = \dfrac{P}{2\pi f}$

The value of τ_{max} for the material which is selected is substituted in the formula

$\tau_{max} = \dfrac{Tc}{I}$. From this we obtain $\dfrac{I}{c} = \dfrac{T}{\tau_{max}}$

Since T and τ_{max} are known, we can obtain a value for $\dfrac{I}{c}$.

In the case of *solid shafts* $\quad I = \dfrac{1}{2}\pi c^4 \qquad$ and $\qquad \dfrac{I}{c} = \dfrac{1}{2}\pi c^3$

What we need to find is the *minimum allowable value* of c. This can be done by

substituting the value of $\dfrac{I}{c}$ which had been found previously into the above

formula. If the *shaft is hollow*, the value $\dfrac{I}{c_2}$ has to be used, where c_2 is the

outer radius of the shaft.

III Engineering materials (Werkstoffe)

1 Properties of materials (Werkstoffeigenschaften)

1.1 Basic concepts (Grundlagen)

The choice of materials to be used for a particular application will depend on *many factors*. The *properties of the materials* are usually the first things to be considered, but other factors like *availability* and *cost* also play an important part. Among the properties which need to be considered, are physical properties, chemical properties and mechanical properties.

Physical properties include quantities like density, electrical resistance, linear expansivity and the melting point. *Chemical properties* of interest are properties like oxidation, combination with other elements and compounds, corrosion and chemical reactions with other substances. Materials should also be recyclable, and must not have an adverse effect on the environment.

Mechanical properties are the ones that are of the greatest importance in the engineering industry. When a material is stressed, it undergoes either an *elastic deformation* or a *plastic deformation* depending on the magnitude of the load. In *structural applications* like columns and machine frames, it is important to keep loads *within the elastic range,* so that *no plastic deformation* takes place. In applications like the *stretching, bending* and *deep drawing* of metal, plastic deformation *has to take place*, if the shape of the metal is to be changed. A brief summary of the most important mechanical properties is given below.

1.2 Some mechanical properties (Einige mechanische Eigenschaften)

- *Strength* refers to the ability of a material to resist tensional or compressional stresses without breaking. The *yield strength*, the *ultimate tensile strength* and the *breaking strength* of a material are important quantities in engineering design.

- *Elasticity* is the ability of a material to return to its original shape and size after the load has been removed.

- *Plasticity* is the ability of a material to be permanently deformed without breaking.

- *Ductility* refers to the ability of a material to undergo *deformation* under *tension* without *rupture* as in a wire or tube drawing operation.

- *Malleability* on the other hand refers to the ability of a material to withstand *compression* without *rupture* as for example in forging or rolling.

- *Toughness* refers to the ability of a metal to withstand the application of *shear stresses* like bending without *fracture*. Copper is by this definition extremely tough, while cast iron is not. Toughness is clearly different from strength.

- *Brittleness* refers to the *ease* which with a *material breaks*. Brittle materials are said to be *fragile* can only be *deformed elastically*. They *break easily* when subjected to *plastic deformation*. Brittleness is the *opposite of toughness*. Cast iron is a brittle material which breaks easily in comparison with steel, which has a much higher breaking strength.

- *Impact properties* – An impact test gives an indication of the *toughness* of a material and its *ability to resist shock*. Brittleness resulting from incorrect heat treatment or other causes *may not be revealed* by a *tensile test*, but is usually shown in an *impact test*.

- *Hardness* refers to the ability of a material to *resist abrasion* or *indentation*. The hardness of a material can be indicated by a hardness test like a Brinell test. Hardness is often a *surface property*.

- *Fatigue* – When a material is subjected to constant loads well below the yield strength of the material, permanent deformation does not normally occur. However the application of *repeated alternating stresses* well below the yield strength can cause cracks to appear. These cracks can gradually become bigger, and finally lead to *fatigue fracture*.

- *Creep* – Tests such as the tensile test and the impact test give information about the behaviour of a material over the *short term*. However, when a metal is loaded over a *long period of time*, it may exhibit gradual extension and ultimately fail even though the stress is *well below* the *ultimate tensile stress*. This phenomenon usually occurs at *high temperatures* and is called creep.

- *Properties which facilitate manufacture* like easy machinability, free flow of the melting metal during casting, and good welding properties also need to be considered when a material is being chosen.
  A product can be manufactured in many ways, for example by casting, by welding, by machining, etc. One has to decide not only on the most appropriate type of material, but also on the *process* which is most *suitable* and *cost-effective* for use with the chosen material.

1.3 The testing of materials (Werkstoffprüfung)

Since properties like ductility, malleability and toughness *cannot be expressed* in *numerical terms*, it has become necessary to use mechanical tests to obtain *numerical values* which help us to compare different materials. Tests such as the tensile test, and the impact test give values which form the *basis of material selection* and *engineering design*. In addition to testing the materials, it is also necessary to carry out tests on the *finished component* to see if it performs satisfactorily under *actual working conditions*, and under possible conditions of overloading. *Long term testing* may be required to avoid failure due to phenomena such as creep and fatigue.

2 Iron and steel (Eisen und Stahl)

2.1 The production of iron (Herstellung von Eisen)

2.1.1 The production of pig iron (Herstellung von Roheisen)

The first step in the manufacture of iron and steel is the production of a very *impure form* of iron called *pig iron*. This is done by reducing iron ore (which is mainly composed of iron oxides) in a *blast furnace*. The addition of materials like *coke* and *limestone* is necessary to carry out this process, and in the end the pig iron together with the *slag* is run off through an outlet near the bottom of the furnace. *Pig iron* is the *raw material* for all iron and steel products. In addition to iron, pig iron contains up to 10% of other elements like carbon, silicon, manganese, phosphorous and sulphur.

2.1.2 The production of wrought iron (Herstellung von Schmiedeeisen)

Wrought iron is formed by removing carbon and other impurities in a puddling (open hearth) furnace. This is a highly *refined form* of *iron* which contains a minute amount of slag. The slag is aligned along the length of the iron and gives it a fibrous structure. Wrought iron is a *tough material* which can be welded with ease and is highly resistant to shock. It can be *bent* and *formed* easily into various shapes and also *does not rust* easily.

2.2 The manufacture of steel (Stahlerzeugung)

Steel is an *alloy of iron* and *carbon*. The carbon content *should not exceed* 2 % if the carbon is to be remain alloyed with the iron. If it exceeds 2 % , the carbon *separates out* as *graphite* and the material is called *cast iron*. Steel is manufactured from pig iron and scrap by oxidizing the impurities in these substances. This is done in various types of furnaces.

2.2.1 The oxygen furnace (Sauerstoff – Blasverfahren)

This is a barrel shaped furnace, which is open at the top and closed at the bottom. Scrap which is sufficient to fill about 30 % of the furnace is put in first, and then molten pig iron is poured in. *Pure oxygen* is blown in *from above* so that it strikes the *surface* of the molten metal. Some of the iron in the melt is converted into *ferrous oxide*. This rapidly reacts with the impurities to remove them from the metal. *Limestone* is added as a *flux*. The process is a very rapid one and is usually completed within half an hour.

2.2.2 Electric steel furnaces (Elektrostahlverfahren)

Electric arc, *induction* and *resistance* furnaces are used to produce high quality steels. The most commonly used furnace is the arc furnace which uses a three phase mains power supply. The furnace is charged with scrap steel, and impurities are *oxidized* by the use of *gaseous oxygen* or *iron oxide*. The slag is removed, and *alloying elements* are added until the *desired composition* is obtained. Induction furnaces are used to make a particular type of alloy steel by first starting with high grade scrap of known composition. During the melting process more elements are added until steel of the right composition is obtained.

2.3 The heat treatment of steels (Die Wärmebehandlung der Stähle)
2.3.1 The basis of heat treatment (Wärmebehandlungsbasis)
Steel is unique in its ability to exist both as a *soft material* which can be *easily machined*, and after heat treatment as a *hard material* out of which *metal cutting tools* can be made. Steel is an alloy of iron and carbon and it is the *presence of carbon* as an *alloying element* which makes these changes in hardness possible.

To understand why the changes in hardness occur, consider what happens to iron when it is heated. When pure iron is heated to 910°C, its crystal structure changes from a *body-centred cubic* structure to a *face-centred cubic* structure. When cooled below 910°C, it *changes back* to the *body-centred cubic* structure. This *reversible transformation* is very important because up to 1.7 % carbon can dissolve in face-centred cubic iron, forming a *solid solution*, while no more than 0.03 % carbon can dissolve in body-centred cubic iron.

The term *alpha iron* is used to denote the body-centred cubic structure of iron existing below 910°C containing the weak solid solution called *ferrite*. The term *gamma iron* denotes the face-centred cubic form of iron containing the stronger solid solution called *austenite*. When the carbon is not in solution in iron, it forms the compound iron carbide (Fe_3C) which is called *cementite*.

Steel containing carbon at room temperature is not hard. The amount of carbon in solution is less than 0.03 %. When steel is heated above the *critical temperature*, the crystal structure changes and up to 1.7 % carbon can dissolve in it. If the steel is now *cooled suddenly*, the carbon does not have sufficient *time to separate* to form a carbide. The sudden cooling of the face-centred cubic austenite leads to the formation of highly stressed *needle-shaped crystals* called *martensite*. The steel becomes extremely *hard* and *brittle* due to the presence of martensite, and is usually subjected to a *tempering process* which reduces both the brittleness and hardness while increasing the toughness.

2.3.2 Stress relieving (Spannungsarmglühen)
Metals become *work (or strain) hardened* when they undergo *plastic deformation* by hammering, rolling, etc. Work hardening occurs because the individual grains of steel become *elongated,* thereby changing the *microstructure* of the steel. Metals in such a state are *under stress* and can *break easily*. A stressed metal component to have its *stress removed* before it can be used. This can be done by a *stress relieving process* in which the components are heated to a temperature of about 600°C for about one or two hours (Fig 2.1). Although the grain structure may not change, this process is popular because it uses a relatively *small amount of energy*. A further advantage is that the surface is not spoilt by *oxidation* or *scaling*. *Recrystallization* and a change in grain structure can take place if the metal is heated for a longer time at this temperature.

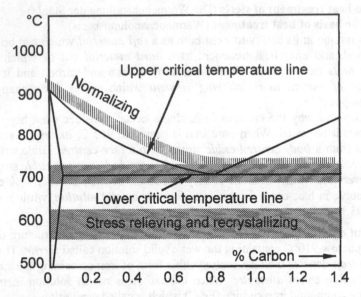

Fig 3.1 Temperatures for the normalizing, annealing and stress relieving of steel

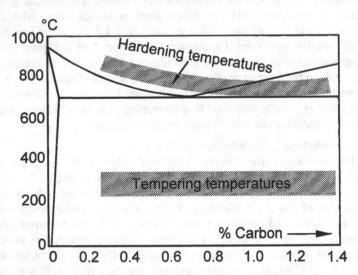

Fig 3.2 Temperatures suitable for the hardening and tempering of steel

2.3.3 Annealing (Weichglühen)

Annealing is a process which makes steel (and also other metals) *completely soft*, while at the same time increasing its *ductility, toughness and machinability*. In the annealing process, steel is first maintained at a temperature of about 700°C for several hours. When the heating is stopped, the furnace and the steel components are allowed to cool down slowly together.

2.3.4 Normalizing (Normalglühen)

Normalizing is a process which is used when the steel has a *coarse grain structure*, or is *nonhomogeneous* and *stressed*. In this process, the steel is heated above the *upper critical temperature* (Fig 3.1) for a short time, and allowed to cool in air at room temperature. This gives the steel a *finer grain structure* which makes it stronger, more homogeneous and stress free.

2.3.5 Hardening (Härten)

There are two reasons for hardening steel components.

*   Steel components are hardened to make them more *wear resistant*.
*   Metal cutting tools made of steel are hardened so that they will acquire the ability *to cut through steel* and other metals which are in a softer state.

Steel is hardened by heating it to a temperature above the *upper critical temperature* (Fig 3.2) and then *quenching* in water or oil. The steel is now *glass hard* because it has acquired a *needle like* microstructure called *martensite*.

2.3.6 Tempering (Anlassen)

Steel that has been *hardened* and *quenched* is *too brittle* and *can break easily*. It has to be *tempered* before it can be used. The tempering process *reduces the hardness* and *brittleness* of the steel while *increasing its toughness*. In the tempering process, the steel component is heated to a suitable temperature *below* the *lower critical temperature* (Fig 3.2) and then allowed to cool slowly.

The temperature to which the component is heated will depend on the *purpose* for which the *component is to be used*. The surface is often polished before tempering. This makes it possible to observe the *colour* of the *oxide film* formed on the surface, which gives an indication of the temperatures reached.

Razors, cutting tools, etc which have to a high degree of hardness but need not be too tough, are tempered at 200 to 250°C. They acquire a pale yellow surface colour. *Screwdrivers, springs, etc* which have to be tough but not so hard, are tempered at a higher temperature of 300 to 350°C and acquire a deep blue surface colour.

2.3.7 Surface hardening (Härtung von Oberflächenschichten)

It is quite often necessary to have a component with a *hard outer surface* and a *tough interior*. If the whole component is hardened, it can become brittle and break. Two of the methods used for *surface* or *local hardening* are given below. These methods can only be used with steels having a *carbon content* of more than 0.35 % . *Mild steel* is *unsuitable* for this purpose.

2.3.8 Flame hardening (Flammhärten)

In this process the surface of the steel is heated by a flame to a temperature above the *upper critical temperature*, so that a hardenable layer is formed on the surface. The heated part is *immediately quenched* (usually by spraying water). The depth of the hardened layer can vary from 1.5 mm to about 6.5 mm.

After quenching, *tempering* to a *temperature* of about *200°C* is usually sufficient. Flame hardening can be used for castings, forgings, etc. Applications include the hardening of machine parts, gear teeth, cams, etc.

2.3.9 Induction hardening (Induktionshärten)

Local hardening can also be achieved by using electrical heating. *Induction heating* is mainly used for the *surface hardening* of cylindrical parts. Coils carrying a high frequency current are placed round the component. Due to the *skin effect*, the *high frequency currents* which are *induced* in the *steel component* to be hardened, *penetrate* only the *surface layers*. The surface layers are heated above the upper critical temperature and become hardened, while the *interior* remains at *a lower temperature* and does not become hard. The surface is quenched by spraying with water.

2.3.10 Case hardening (Einsatzhärten)

Components having a hard surface and a soft interior can also be produced by case hardening. There are two stages in this process (a) carburizing, (b) heat treatment, which includes *hardening tempering and refining*. Low carbon steel with 0.1% to 0.2 % carbon is used at the start.

(a) **Carburizing** (Aufkohlen) – In the carburizing process carbon is made to *diffuse* into the *surface of low carbon steel* at high temperatures so that the carbon content of the outer layers exceed 0.35 % , while the carbon content of the interior remains lower at about 0.2 %. The outer layers can be hardened while the interior cannot be hardened. Carburizing methods can be divided into *three groups*.

Solid carburizing is carried out by using a mixture of sodium, barium and other carbonates and wood charcoal. The components and the carburizing mixture are packed into boxes made of heat resistant alloy. The boxes are slowly heated to about 925°C, and are kept at this temperature for several hours. Case depths (or hardened layers) of 1 mm or more can be achieved.

Liquid carburizing is done in baths containing sodium cyanide together with sodium or barium chloride. The baths are heated to about 900°C and the components are immersed for periods of up to 2 hours depending on the case depth required. *Continuous operation* and *automatic quenching* are possible keeping *operating costs low*.

Gas carburizing is carried out in both batch type and continuous furnaces. The components are heated for 3 to 4 hours to a temperature of 925°C in an atmosphere of gases like methane, ethane or carbon monoxide. All these gases break down, depositing carbon on the exposed surfaces. Here again *automatic quenching* and *low labour costs* are possible.

(b)Heat treatment after carburizing (Wärmebehandlung der Einsatzstähle)
If the carburizing has been done correctly, the core will have a low carbon content (0.1 to 0.2 % of carbon) while the case (outer layers) will have a maximum content of 0.83 % . The prolonged heating produces a *coarse grain structure in the core,* and the size of the grains are reduced by a process called *refining.* This increases the toughness and the strength of the core.

Refining the core – This is done by heating the component to just above its upper critical temperature (about 870°C for the core) when the coarse ferrite/pearlite structure is replace by refined austenite crystals. The component is next quenched in water so that a fine ferrite/martensite structure is obtained.

Refining the case (or surface layers) – The component is heated to about 760°C at which temperature the coarse martensite of the case changes to fine grained austenite. Quenching produces a fine grained martensite in the case.

Present day cost saving procedures – The above procedure is costly. At the present time it has become possible to adopt the cheaper procedure of *quenching* the component *direct from the carburizing medium.* This is followed by *tempering* at about 200°C to remove the quenching strains. This procedure works satisfactorily due to the *availability of steels* which *maintain a fine grain size* at *carburizing temperatures.* Distortion of the component is less than that experienced in other types of case hardening.

2.3.11 Nitriding (Nitrieren)

In the nitriding process, special alloy steels containing aluminium, chromium and vanadium are heated to temperatures of about 500°C for 50 to 90 hours in a chamber through which ammonia gas is passed. *Hard nitrides* of the alloying metals as well as iron are *formed on the surface.* All machining and heat treatment operations must be carried out before nitriding.

Advantages of nitriding are that the hardness is maintained at *high temperatures* and that the *fatigue resistance* is also increased. *Resistance to corrosion* is good if the component is left unpolished. The process is cheap if large numbers of components are treated simultaneously.

2.4 Steel types (Stahlsorten)

A very large number of steel types are presently available being manufactured on accordance with *specifications* set by various *private* and *governmental institutions.* In the U.S.A. the relevant bodies are the American Society for Testing and Materials (ASTM), the Society of Automotive Engineers (SAE), The American Society of Automotive Engineers (SAE). In Europe the EU specifications are relevant, while in Germany the DIN specifications are used.

Steel is basically an alloy of iron and carbon, and steel that contains only these two elements is called *carbon steel.* Low carbon steels with a carbon content of up to 0.25 % (also known as *mild steel),* are used where *moderate strength* with *high plasticity* is called for. These steels *cannot be hardened* or heat treated.

Medium carbon steels having up to 0.55 % carbon are called *machinery steels* and can be heat treated to *acquire strength*. *High carbon steels* contain from 0.60 to 1.30 % carbon can be hardened and are used to *make tools*.

2.4.1 General purpose steels (Allgemeine Stähle)

These include *carbon steels* and steels with a *low content* of *alloying metals*. These steels which are very widely used have varying compositions and heat treatment requirements which are tailored to the application. They must have good mechanical properties like strength, toughness, ductility, weldability, etc. Applications include steel for buildings, containers, vehicles, machines, etc. *Rolled sheet* and *drawn wire* are the starting points for numerous products (see p 193) like cans, containers, vehicles, screws, bolts and nuts, etc.

2.4.2 Fine grained welding steels (Schweißgeeignete Feinkornstähle)

These are special steels with a carbon content of less than 0.22 %. Alloying metals like aluminium, niobium and vanadium are added to produce a steel that has the properties of *high strength* and *good weldability*.

2.4.3 Quenching and tempering steels (Vergütungsstähle)

These are steels with a carbon content of 0.2 to 0.65 % and are available as plain carbon steels or steels with a low alloy content. They need to be *heat treated* (hardened and tempered) *before use* because heat treatment increases their strength and toughness.

2.4.4 Case hardening steels (Einsatzstähle)

These are steels with a *low carbon content* (about 0.20 %) and are available as plain carbon steels or steels with a low alloy content. They are used to manufacture components with a hard surface and a tough interior. This is achieved by case hardening.

2.4.5 Free-cutting steels (Automatenstähle)

These steels are used in *automatic lathes* for the mass production of small components. They are plain carbon steels in which the *sulphur* and *phosphorous* content have been increased. *Lead* is also frequently added to the steel. The addition of these elements result in a steel in which chips produced during the turning process are short and do not break easily. This results in an *improved surface finish*.

2.4.6 Nitriding steels (Nitrierstähle)

These are special alloy steels which contain *nitridable elements* like chromium, aluminium and titanium. Components which have been nitrided have a hard surface and a tough interior.

2.4.7 Spring steels (Federstähle)

Spring steels which are used in machines and vehicles are usually plain carbon steels of high purity. For other applications, alloy steels which can be hardened and tempered are used. Spring steels must in general have a *high elastic limit*, *high tensile strength* and good *resistance to fatigue*.

2.4.8 Stainless steels (Nichtrostende Stähle)
These are steels that have the ability to resist *rusting* and *corrosion*. They have a chromium content of at least 12% and in addition other metals like nickel, vanadium and molybdenum.
They are used in the chemical industry for containers, pipes, machine parts,etc. They are also used in the food industry and in domestic kitchens for cooking utensils, sinks, dishes, etc.

2.4.9 Low temperature steels (Kaltzähe Stähle)
These are steels that *do not become brittle* and *retain* their *toughness* at temperatures below −50°C. They are used for making containers, pipes and other items which are required for the production and transport of liquid gas.

2.4.10 Heat resistant steels (Warmfeste Stähle)
These steels need to retain *good mechanical properties* at temperatures of up to 800°C. Plain carbon steels can be used up to a temperature of about 400°C and steels with a low alloy content of metals like chromium, molybdenum and vanadium can be used up to about 540°C. For higher temperatures, steels containing up to 12 % chromium with other metals like nickel, molybdenum, titanium and vanadium are needed. Heat resistant steels are used in the chemical industry and in power plants.

2.4.11 Corrosion resistant steels (Korrosionsbeständige Stähle)
These are special steels that must have a resistance to the corrosive effect of gases at temperatures of about 500°C. In these steels the *chromium content is increased* by up to 30 %.

2.4.12 Steel for electrical machines (Stähle für elektrische Maschinen)
Steels used for electrical work must have special magnetic properties. Two types of steel are required.
a) **Soft magnetic materials** are steels which *lose their induced magnetism* when the *magnetizing field* is *removed*. These materials are used in the *magnetic cores* of transformers, motors, generators etc., where an alternating magnetic field has to exist.
   Soft magnetic materials used in AC machines are *iron-silicon alloys* with practically no carbon. The metal is supplied as sheets from which *laminations* are stamped. The use of laminations (thin sheets) in the core of machines and transformers *reduces eddy current losses*.
   Transformers and coils used in the communication and audio industry need materials with a *high permeability* and *low hysterisis loss*. Alloys for this purpose have a high nickel content and also other metals like cobalt.
b) **Hard magnetic materials** are steels which *retain* their *magnetism* when the *magnetizing field* has been *removed*. Hard magnetic materials are used in *permanent magnets*.

A permanent magnet should *retain* its *magnetism* for a long time (after the magnetizing field has been removed). The flux in a permanent magnet depends on its design and on the steel alloy from which it is made. The steels used for permanent magnets nowadays are aluminium nickel alloy steels.

2.4.13 Tool steels (Werkzeugstähle)

Tool steels are used in tools for operations like shearing, cutting, forming, extruding, rolling, etc. *Cutting tools* should be *hard* and *resistant to wear*. They should have the ability to *cut at high speeds* and *high temperatures*. *Forming tools* should be *tough* and *shock resistant*. Tools used for tasks like *extruding* or *hot rolling* should be able to maintain their *strength* at *high temperatures*. Tool steels may be divided into the following categories.

a) **Cold working steels** may be used in applications where the surface temperature does not exceed 200°C. They are available as plain carbon steels and also as alloy steels.

b) **Hot working steels** are able to work continuously at temperatures of up to 500°C. They are used for making *extrusion* and *casting dies* as well as in press applications like *blanking* and *forming*. They must possess good mechanical properties at high temperatures like toughness, strength, wear resistance and the ability to withstand deformation. These steels contain alloying elements like tungsten, molybdenum, vanadium and chromium.

c) **High speed tool steels** – These steels are used to make *metal cutting tools* which are able to work at *high speeds* and *temperatures* of *up to 600°C* without losing their hardness. They have a high carbon content of above 0.75 % and contain alloying metals like tungsten, molybdenum, chromium, vanadium and cobalt.

d) **Special purpose steels** – There are steels used for special purposes like dies for *die casting*, dies for the *hot pressing* of *nonmetals,* plastic forming, etc.

2.4.14 Steel castings (Stahlguß)

Steel castings have the *advantages* that are usually *associated with steel,* like good tensile strength, toughness, high rigidity, good welding properties and excellent endurance properties. Steel castings with a high alloy content have good heat and corrosion properties. Compared with cast iron, the *higher melting temperatures* and poorer *flow properties* during casting are *disadvantages*. In the manufacture of complex parts, cast steel parts have a considerable *cost advantage* over forged parts. Small complex parts can be made by precision casting methods like *investment casting* (lost wax casting). Most of the castings are from plain carbon or low alloy steel. Steel castings are used for *mechanically stressed* parts for turbines, valves, aircraft engine parts, machine components, etc. Steel castings are also used in the construction of railroad cars, trucks, trailers, tractors, etc. Large castings find use in ships, rolling mills, and in the mining and logging industries.

2.5 Cast iron (Gußeisen)

2.5.1 General properties (Allgemeine Eigenschaften)

The term cast iron has been used for iron with a *carbon content* of between 2 and 4.5 %. Cast iron usually contains varying amounts of *impurities* like silicon, sulphur, manganese and phosphorous. *Alloy cast irons* contain alloying elements like nickel, chromium, vanadium, etc. Cast iron is made in furnaces from a charge of pig iron, steel scrap, coke, and scrap from castings. In general there are two types of cast iron, *white cast iron* and *grey cast iron.*

1. **White cast iron** is produced by the rapid cooling of the melted charge. The carbon is present as *iron carbide* (cementite). Cementite is a hard, white, brittle compound, and cast irons containing this show a *white fractured surface* when broken.

2. **Grey cast iron** is produced when the cast iron is cooled slowly. The carbon exists in the form of *graphite*. A fractured *surface appears grey* and such irons are called grey irons.

Silicon dissolves in the ferrite of a cast iron and has a *controlling influence* on the relative amounts of graphite and cementite present. When the silicon content is small, the carbon is in the form of cementite. Increasing the silicon content causes decomposition of the cementite, producing graphite and grey cast iron.

Cast iron is a relatively *cheap material* compared to steel. It can be cast at a lower temperature than steel and is a *hard wearing material*. However it is *brittle* and *lacks toughness*. It has good corrosion resistance.

2.5.2 Grey or lamellar cast iron (Gußeisen mit Lammellengraphit)

Grey cast iron is the most *frequently used* of all cast irons. Carbon is present in the form of *graphite flakes*. The tensile strength of the cast iron depends on the size of the flakes, being higher when the flakes are finer in composition. The compression strength is four times higher than the tensile strength. The malleability and impact strength of this type of cast iron is poor. Grey cast iron has *good damping* and *sliding properties*.

The presence of graphite makes the iron softer and gives it *good machining properties*. It is used for making machine beds because of its *good damping properties*. If small quantities of alloying elements like chromium, nickel, molybdenum and copper are added the *strength* can be *increased*. Grey cast iron is used to make engine blocks, cylinder heads, machine frames, etc.

Meehanite is a cast iron with particularly *fine graphite flakes*. It has good strength and is free of stress. It is *free of defects*, holes, etc. and can be *hardened* and improved by heat treatment.

2.5.3 Nodular (ductile) cast iron (Gußeisen mit Kugelgraphit)

The long thin flakes of graphite in grey cast iron have negligible tensile strength and act as *discontinuities* in the structure. A great improvement can be achieved if the graphite flakes are replaced by *spherical particles of graphite.* The formation of spherical (or nodular) graphite is effected by adding small quantities of *cerium* or *magnesium* (up to 0.5 %) to the molten iron.

Magnesium is added in the form of a nickel-magnesium alloy. Nodular cast iron is *stronger and tougher* than grey cast iron. A further improvement in mechanical properties is possible by *alloying* and *heat treating*. Nodular cast iron is used for crankshafts, gear wheels, machine housings, pipes for the chemical industry, etc.

2.5.4 Malleable cast iron (Temperguß)

Malleable cast iron is used to make components which have *complicated shapes* and also need to be *tough* and *impact resistant*. The castings are brittle when cast, but have their malleability increased considerably by subsequent *heat treatment*. There are two types of malleable cast iron, whiteheart cast iron and blackheart cast iron. The original castings used as starting points in both these processes are made from brittle white cast iron.

Whiteheart process – In this process, the castings are heated for days at about 1050°C in an oxygen rich *decarbonizing atmosphere*. The carbon at the surface is removed by oxidation and the outer layers are ferritic in character. When fractured they present a white appearance and hence the name whiteheart. In castings which have larger cross-sections, the interior can contain some graphite.

Blackheart process – In this process, the castings are heated in a neutral atmosphere for about 30 hours at about 950°C. This results in the breakdown of the cementite into ferrite and graphite in the form of small rosettes of temper carbon. A fractured section appears black and hence the term blackheart.

2.5.5 Hard cast iron (Hartguß)

This is a cast iron in which there is no graphite and the carbon exists entirely in the form of cementite. Hard cast iron is able to withstand compressional stress, but not tensional stress. It is *extremely brittle*. By adding alloying elements and by subjecting the material to suitable cooling processes, it is possible to have castings with a hard surface and a tough interior. This material is used for castings which need to be *hard* and *wear-resistant* like rollers, camshafts and deep drawing tools.

3 Nonferrous metals (Nichteisenmetalle)

3.1 Copper and its alloys (Kupfer und Kupferlegierungen)

Copper is a metal with a *high electrical conductivity* and is used in electrical cables. The extremely pure copper required for this is refined electrically. Copper is *soft, ductile* and *very tough*. Although copper is corrosion resistant, a green layer of *copper carbonate* called *verdigris* forms on the surface when it is exposed to the atmosphere. Copper can be cast, soldered or welded without any difficulty. It is used as a conductor for power and other cables. It is also used in applications where a *high thermal conductivity* is required. Large quantities of copper are used in the making of alloys like brass and bronze.

3.1.1 Brass (Messing) – This is a Cu-Zn alloy and is the *most used* of the *heavy nonferrous metal alloys*. The copper content must at least be 50 % , otherwise the alloy becomes too brittle.

The yellow colour, the high degree of polish, and the resistance to corrosion make brass a suitable alloy for decorative purposes. It is used for watch parts, electrical parts, pinions, hinges, brackets, etc.

High tensile strength brass (Messing mit höher Zugfestigkeit) – This is a brass with good *hot working* properties, *high tensile strength* and *resistance to abrasion*. It is sometimes called *manganese bronze* due to an oxidized bronze appearance on the surface of extruded parts.

Free-cutting brasses (Einfach zerspanbarer Messing) - These are of the 60-40 type and contain about 2 % of lead. They are *easy to machine*.

3.1.2 Bronze (Zinnbronze)

The bronzes are copper-tin alloys containing 83 to 98 % copper, and 2 to 15 % tin. Some of them also contain other metals like zinc, nickel and lead.

Tin bronzes – These bronzes combine *hardness* and *ductility* with *high resistance to corrosion*. Wrought alloys have a tin content of up to 10 %. Cast alloys have a tin content of up to 20 %. Due to the good *sliding properties* and the *resistance to wear*, they are used to make highly stressed shell bearings and worm gears.

Phosphor bronzes – Phosphor bronzes contain between 0.1 and 1 % of phosphorous. The phosphorous is supposed to increase the *tensile strength* and the *corrosion resistance*. Phosphor bronzes are mainly used for plain bearings and other components where a *low friction coefficient* combined with *high strength* are required.

Bronzes containing zinc – These bronzes are available as *wrought* and *cast* alloys. The wrought alloys were mainly used to make coins, while the cast alloys were used where strong *corrosion resistant castings* were required.

Bronzes containing lead – These bronzes contain at least 60 % copper and a lead content of up to 35 % . In addition they contain alloying metals like tin, zinc and nickel. Since lead is not soluble in copper, it appears in the form of spheres in the bronze. This results in good *lubricational properties* allowing *higher loading* and *higher speeds* for bearings made from this type of bronze. These are used for aero and automobile crankshaft bearings.

Aluminium bronzes - Aluminium bronzes are available as wrought and cast alloys with an aluminium content of up to 11 % . These alloys are able to *retain* their *strength* at *elevated temperatures*, have a *high resistance to oxidation* and good *corrosion resistance* at ordinary temperatures. They have good wearing properties and some alloys have a pleasing colour which makes them a *substitute for gold* in imitation jewellery. Their high resistance to corrosion which results from an exposure to salt water, make them useful for *ship propellers* and *turbine blades* as also for parts in the chemical industry. These alloys are difficult to solder or weld.

3.1.3 Copper-Nickel alloys (Kupfer-Nickel Legierungen)

These alloys contain 40 to 45 % of nickel. They are the *most corrosion resistant* of all copper alloys. Alloys called *nickel-silver* contain 50-63 % of copper, 10 to 23 % of nickel and the remainder zinc. They are white in colour and have been used to make *tableware* and *ornamental objects*. They are also used for making coins, for resistance wire in the electrical industry, and in the chemical industry.

3.2 Nickel and nickel alloys (Nickel und Nickel Legierungen)

Nickel is a metal with a *high tensile strength* and a *high degree of toughness.* The addition of manganese increases the strength without affecting the toughness. It retains its strength at 500°C and remains tough at low temperatures. Nickel can be cast, welded and soldered without difficulty.

Alloys like *Monel* (nickel 67 % , copper 30 %, the remainder iron and manganese) are used because of their good high temperature properties and their resistance to corrosion, in the manufacture of valves, blades for turbines, equipment for chemical plant, etc. Nickel alloy (Nichrome) wires are used in the manufacture of precision *electrical resistors*.

3.3 Zinc and its alloys (Zink und Zink Legierungen)

Zinc is a cheap metal which is very much used as a *protective coating* for steel which is exposed to the atmosphere. A coating of zinc carbonate is formed on the surface which protects the metal from further erosion. The process known as *hot dip galvanizing* is widely used to protect all kinds of steel objects and structures. Zinc shows *poor corrosion resistance* against *acids and salts.*

Zinc alloys which contain aluminium and copper are used in *die casting processes*. The low melting point of these alloys is an advantage. Objects made by the die casting process have excellent *dimensional accuracy* and *surface finish* eliminating the need for further finishing processes. However these objects lack stability at high temperatures and have poor *corrosion resistance.*

3.4 Tin and its alloys (Zinn und Zinn Legierungen)

Very pure tin is used largely as a *protective coating* for steel to make *tinplate,* which is used to make containers in the food industry. It is used in industry as an alloy with lead to make *solder. Pewter* is an alloy used for ornamental objects and consists 91 to 93 % tin, 6 to 7 % antimony and 1 to 2 % copper.

3.5 Lead and its alloys (Blei und Blei Legierungen)

Pure lead is used in the chemical industry because of its *corrosion resistance* particularly against *sulphuric acid*. It is used as a *screening metal* i.e., to protect people and objects from the harmful effects of X′rays and radiations from radioactive substances. It can be easily cast, welded or pressed and is used for making plates for accumulators. Telecommunication and other cables which are laid underground usually have a *protective coating* of lead.

3.6 Aluminium and its alloys (Aluminium und Aluminium Legierungen)
Aluminium is a silver white metal with a density of about *one-third the density of steel*. When exposed to the atmosphere an *oxide coating* forms on the surface and this coating prevents further oxidation. Pure aluminium is relatively *soft* and *weak* and for engineering applications aluminium is mostly used in the form of an alloy. Aluminium alloys containing small amounts of other elements are used to make castings for the aero, automobile and constructional industries.

Alloying and heat treatment can produce aluminium components which are weight for weight *stronger than steel* and this fact has lead to the extensive use of aluminium alloys in *air-frame construction*.

Aluminium is a good conductor of both heat and electricity. It is ductile and is particularly suitable for manufacturing objects by cold drawing and cold pressing. The strength of aluminium is increased by *work hardening* processes like rolling, drawing, pressing and hammering.

3.6.1 Aluminium alloys (Aluminium Legierungen)
The addition of alloying elements is carried out mainly to improve the mechanical properties like tensile strength, rigidity, hardness and machinability. Sometimes alloying improves casting properties like fluidity. The chief alloying elements are copper, magnesium, manganese, zinc and nickel.

3.6.2 Wrought alloys of aluminium (Knetlegierungen)
Some wrought alloys are meant to be heat treated and some are not. The ones that are meant to be heat treated contain magnesium and silicon and are *resistant* to sea water *corrosion*.

Heat treatment improves the *strength of the alloy* without impairing its ductility and cold formability. Heat treatment is carried out by first heating the alloy components in an oven or salt bath to temperatures of about 500°C which is close to their melting point. The objects are then quenched in water and then allowed to remain at room temperature for a period of days. The strength is increased in this way by a process termed *age hardening*. This process *can be shortened* by keeping the quenched alloy at higher temperatures of 100 to 200°C.

3.6.3 Cast aluminium alloys (Aluminium Gußlegierungen)
These alloys are used in sand castings and also in gravity and pressure die castings. The most important alloys are those containing about 12 % silicon and of approximate eutectic composition. These alloys have good fluidity. The coarse eutectic structure can be changed to *fine grained structure* by adding small amounts of sodium just before casting. Silicon alloys have *high strength* while the addition of magnesium gives good *corrosion resistance* and *heat conductivity*. The aluminium-copper-titanium alloy castings have the *highest strength* and are used in components for aircraft and automobiles provided the castings have a microstructure which is free of failures. Pressure die castings have good dimensional accuracy and surface quality. However oxidation leads to defects, bubbles, etc. in the castings, which reduce their strength.

3.6.4 Anodizing of aluminium (Anodisieren von Aluminium)
Anodizing is a way of improving the *corrosion resistance* of aluminium and its alloys. It also produces a *coloured finish* on the surface which may be desirable for ornamental purposes. In the anodizing process, the parts to be anodized are made the anode in an electrolytic bath containing chromic, oxalic or sulphuric acid. When a current is passed through the circuit, a tough coating of aluminium oxide is formed on the surface of the parts. To obtain a coloured finish, the parts are dyed by immersion in cold baths of *dyestuff*. The *porous* anodic coating can be *sealed* by treating it with hot or boiling water.

3.6.5 Uses of aluminium (Verwendung von Aluminium)
Aluminium is *second only to steel* in its *usefulness*. It is used in transport vehicles like aircraft, trains, ships etc. It is also used in packing (films, tubes, boxes, cans, containers etc.), in electrical work (cables, capacitors, wiring, switches, light housings, etc.), in household items (plates, vessels, containers, etc., and in numerous other objects like instruments and machines.

3.7 Magnesium and its alloys (Magnesium und Magnesium Legierungen)
Magnesium is the *lightest industrial metal* available and is almost never used in a pure form because it *burns easily*. It can however be used in alloyed form, and its alloys are used where *moderate strength* and *extreme lightness* are required. Magnesium alloys are considerably lighter than aluminium alloys and magnesium alloy castings are used in the automobile and other industries for housings, cylinder heads, machinery parts, etc.
Magnesium alloys are available both as wrought alloys and cast alloys. A large variety of castings, forgings and extruded shapes are available for a wide variety of applications. Articles made from magnesium alloys may be joined by welding or riveting.

3.8 Titanium and its alloys (Titan und Titan Legierungen)
Titanium and its alloys have *strengths* that are close to *alloy steels* while their weight is only about *60 % of the weight of steel*. These materials have excellent *corrosion resistance* properties comparable to or even better than those of stainless steel. Titanium has good *fatigue resistance* and a *high melting point*. Its ability to retain its *strength* at *high temperatures* is a property which favours its use in jet engines. Titanium is mainly used in the *aircraft industry* where its strength, toughness and corrosion resistance can be used to advantage. It is also used in the chemical industry for the manufacture of pressure vessels, pumps, cooling pipes, etc.

3.9 Soldering alloys (Werkstoffe für Lötungen)
Solders are nonferrous metal alloys which are commonly used to *join metals*. Solder melts at temperatures that are *lower than the melting points* of the metals being soldered.
Soft solder is an alloy consisting typically of 50 % lead and 50 % tin. **Brazing** or *hard soldering* is similar to soldering, but uses alloys which have a higher melting point. It is used where a *tougher, stronger joint* is required. Brazing

solder which is called *spelt* or *silver solder* contains typically 50 % copper and 50 % zinc.

3.10 Bearing metals (Lagerwerkstoffe)

A bearing metal has to be *tough* and *ductile* so that it can withstand *mechanical shock*, but at the same time *hard* and *abrasion resistant* so that it can withstand *wear*. It must also have have *low friction losses*.

3.10.1 Copper based bearing alloys (Lagerwerkstoffe auf Kupfer Basis)

Plain tin bronzes containing from 10 to 15 % tin and phosphor-bronzes containing 0.3 to 1.0 % phosphorous are widely used where the loads are heavy. For small bearings, *sintered bronzes* are often used, and are made by sintering copper and tin powder together with graphite. These bearings are usually of the *self-lubricating type*.

Leaded bearings are used in the manufacture of *main bearings* for *aero-engines* as also for automobile and diesel *crankshaft bearings*. They are *wear-resistant* and their *good thermal conductivity* keeps them cool while running.

3.10.2 White metal bearing alloys (Lagerwerkstoffe auf Blei oder Zinn-Basis)

These may be either lead base or tin base alloys. These are cast to form bearing surfaces on bronze, steel or cast iron shells. These bearings work satisfactorily against a *soft steel shaft*. At higher temperatures, they are subject to spreading, fatigue and a lowering of strength.

3.11 Precious metals (Edelmetalle)

Gold is used for the manufacture of *jewellery* and other decorative objects, and also in the electrical industry as a *protective coating*, which is deposited by electrical means. Its use as a monetary standard has been declining in recent years.

Silver is used in the manufacture of *jewellery* as well as in the electrical industry for the manufacture of *heavy electrical contacts*. It is also used in the manufacture of *tableware* and in the making of *photographic emulsions*.

Platinum is a metal known for its *chemical inertness, high melting point* and its usefulness in *catalytic reactions*. Its chemical inertness is used to advantage in the manufacture of laboratory equipment. It can be used at high temperatures without a protective atmosphere. It is used in high temperature *thermocouples*, for making *electrical contacts* and in the manufacture of *jewellery*.

3.12 Sinter materials (Sinterwerkstoffe)

In sintering processes (also called powder metallurgy), very fine powders of metals, alloys, refractory materials and compounds are compacted into the final product. The following steps are involved.

1. The powder is produced by *mechanical means* and may be composed of a single material or a mixture of materials.
2. The powder is *pressed* into the desired shape by using suitable moulds.
3. The pressed articles are subjected to a *sintering process* which is essentially a *heating operation*.

Sintering processes have the following advantages:

a) Metals like tungsten which are *difficult to melt*, may be *powdered* and *sintered*. A combination of dissimilar materials and refractory materials may also be sintered.

b) It is a *cost-effective method* for producing parts in *large numbers* even allowing for the high cost of dies.

c) No further *finishing processes* are required. The desired mechanical properties may be obtained by using the *right mix of powders*.

Disadvantages are that *high pressures* and *expensive dies* are required. The size of the components that can be made are limited, and it is *not possible* to make certain forms like *screw threads*.

3.13 Cemented carbides (Sinterhartmetallkarbide)

Cemented carbides are used as tips of *cutting tools* for lathes and other machines. They are *second* only to *diamond* in hardness. They are also used in any other devices where *wear-resistance* is essential, like in wire drawing dies, gauges, die linings, etc. These are produced by using powder metallurgy techniques.

Carbides of metal such as tungsten or titanium are produced by adding carbon to the metal or oxide (or to a mixture of metals and oxides) and heating in a reducing atmosphere to a temperature of about 1400°C.The carbide powder is mixed with a powdered metal binder (usually cobalt), and then pressed and sintered. The strength and hardness can be controlled by varying the quantity of binder. Lathe cutting tools are made by brazing *cemented carbide tips* of the right shape to the steel body of a normal lathe tool.

4 Nonmetallic materials (Nichtmetallische Werkstoffe)

4.1 Ceramic materials (Keramische Stoffe)

4.1.1 Bricks (Ziegelsteine)

Bricks are small blocks of material used primarily for building purposes. They are machine made in moulds under pressure from mixes of clay. The wet bricks from the mould are first dried in air and then baked in ovens at temperatures between 900°C and 1300°C. Bricks are resistant to freezing and attacks from chemicals. They are *resistant to fire* and provide good *thermal insulation* as well as *insulation* against *noise*.

4.1.2 Fire-resistant materials (Feuerfeste Steine)

Fire-clay bricks are used to line *kilns, ovens, furnaces*, etc. They are made out of a mixture of flint and clay and can withstand *temperatures of over 1500°C.*

High-alumina bricks are made from materials like bauxite and diaspore which are rich in alumina. They are used where the *temperature* and *load conditions* are *particularly severe.*

Silica bricks are made from crushed rock which contains up to 98 % silica. As material for bonding 2 % lime is used. The bricks are useful under conditions where the *strength* at *high temperatures* has to be good.

4.1.3 Cement (Zement)

The normal cement used for building purposes is called **portland cement**. It is made from a mixture containing 80 % calcium carbonate (chalk, limestone, etc.) and about 20 % clay. The mixture is finely ground and calcined in kilns to a clinker. After cooling, this clinker is ground to a fine powder. During the grinding process, a small amount of gypsum is usually added. The gypsum regulates the setting of the cement.

4.1.4 Concrete (Beton)

Concrete is made from a **mixture of cement** and a **combination of inert particles** of various sizes like gravel, sand, broken stone etc. When mixed with a suitable quantity of water and placed in moulds, it hardens into blocks having the desired shape.

4.1.5 Reinforced concrete (Stahlbeton)

Ordinary concrete has a **high compressive strength**, but **poor tensile strength**. Reinforced concrete is a **composite material** in which mild steel usually in the forms of bars is imbedded in the concrete. The presence of steel increases the tensile strength of the concrete and makes it possible to use it in beams, pillars, etc. which can **bend without breaking**.

4.1.6 Prestressed concrete (Spannbeton)

This is a form of reinforced concrete where the reinforcing is done by imbedding bars of **high tensile strength** steel. The bars are given **initial stresses** opposite to those caused by the load. This is called **pretensioning**.

4.2 Glass (Glas)

Glass is a **noncrystalline material** which is made by melting a mixture of silica, alkali and stabilizing substances like alumina, lime, lead and barium. Small quantities of manganese and selenium oxide are added to obtain colourless glass. Different types like window glass, laboratory glass, optical glass and also coloured glasses can be made by adding different metallic oxides to the mixture. Glass in molten form can be **moulded** or **fabricated** into different shapes. It is rigid at room temperature, but may be **remelted** and **remoulded** repeatedly. Glass has many uses. It is used for making containers like bottles and also as window glass. Glass fibre is used as a **thermal insulating material** (fibre glass) and also in **glass fibre** telecommunication **cables**. It is used in the building industry in the form of glass bricks and for ornamental purposes.

4.3 Wood (Holz)

Wood is popular as a raw material because of its **strength** and **the ease with which it can be worked**. Its **attractive appearance** makes it a much sought after material in the manufacture of furniture, doors, door ways, ship interiors, etc. Wood is a **fibrous composite material** composed of cellulose, lignin and resins. **Plywood** is made by glueing together an odd number of layers of veneer with **alternate layers** having their grain at **right angles to** each other. The alternation of the direction of the grain tends to give the plywood equal strength in the two face directions.

4.4 Plastics (Kunststoffe)

The term plastic refers to *artificially made organic substances* which do not exist in nature. A feature that is common to all plastics is the fact that they are composed of long *chain-like macromolecules*.

The oldest plastics like celluloid were made from natural materials. More recently plastics have been made from coal, acetylene and mineral oil. Plastics are very much used in the mass production of consumer products, because they are *light, cheap* and *easy to manufacture*. Plastics can be classified into three groups: thermoplastics, thermosetting plastics, and elastomers.

4.4.1 Thermoplastics (Thermoplaste)

Thermoplastics become *soft when heated* and can be moulded into different shapes. They become *hard on cooling*, but can be *reheated* and *remoulded repeatedly*. Larger quantities of mass produced goods are made from thermoplastics as from any other type of plastic. This is due to the ease with which they can be moulded and also due to the large range of plastics available. Some of the most commonly used types are briefly mentioned below.

Polyamide (PA) (Polyamide) – This is milk white in appearance and has a surface with good *sliding properties*. It has a *high tensile strength* and is also *wear-resistant, hard* and *tough*. It is used to make gear wheels, cams, guide pulleys, fuel tanks, protective helmets, etc.

Polyethylene (PE) (Polyethylen) – Also known as polythene, this plastic has a colourless to milky appearance and a *wax-like* smooth *surface*. It is *resistant to acids* and *alkalies*. There are two types as follows:

  Hard polyethylene — poor flexibility, used to make containers and tubes.
  Soft polyethylene — good flexibility, used for films, hoses, wire coatings.

Polymethylmethacrylate (PMMA) (Polymethylmethacrylat) – Known under trade names like acrylglass or plexiglass. It is colourless, glass clear and has a *glossy surface*. It is *hard, tough* and *difficult to break. Resists* the action of *acids* and *alkalies*, but is *soluble* in *certain solvents*. It is used to make optical lenses, safety glasses, transparent housings and roofings, sanitary articles, etc.

Polypropylene (PP) (Polypropylen) – Polypropylene is similar in appearance and has similar properties to hard polyethylene. It is however harder, and retains its shape better at temperatures of up to 130°C, and is therefore able to *withstand boiling water* indefinitely. It is used for washing machine and automobile parts.

Polystyrene (PS) (Polystyrol) – Polystyrene is colourless, has a glossy surface and is glass clear. It is *hard* and *brittle* and *breaks easily*. It is able to withstand acids and alkalies, but shows poor resistance to organic solvents. It can be made less brittle by adding acrylnitrile. It is used to make show windows, containers, glasses, etc.

Polyvinylchloride (PVC) (Polyvinylchlorid) – Polyvinylchloride is transparent and resistant to both acids and alkalies. Two types are available, hard and soft.

The hard PVC is *tough* and *difficult to break*. By adding suitable softeners, the material can be made to have properties similar to *rubber, and leather*. Hard PVC is used to make objects like housings, tubes, valves, etc. Soft PVC is used to make artificial leather, gloves, soles for shoes, boots, etc.

Polytetrafluoroethylene(PTFE) (Polytetrafluorethylen) – Also known as *teflon*, this plastic has a milky white appearance and a wax like slippery *low friction surface*. It is *soft, flexible, tough* and *wear-resistant*. It can be used over a wide temperature range from -150°C to 280°C. It is used to make gaskets, non-stick surfaces, electrical insulation, coatings, lubricants, etc.

4.4.2 Thermosetting plastics (Duroplaste)

Thermosetting plastics can be *heated* and *moulded only once*, and do not become soft when reheated. They are known for their strength, dimensional and thermal stability, resistance to chemicals, durability and good electrical properties. The polymer is mixed with *fillers* before moulding. Fillers include powdered minerals, wood flour, clays, cellulose, glass and textile fibers.

Phenol resin (PF) (Phenolharz) – It is yellow-brown in colour, *hard, brittle* and *fractures* easily. The resin is mixed with fillers before use.

Melamine resins (MF) (Melaminharz) – It is colourless to light yellow, *hard,* and *brittle*. It *fractures* easily. In a pure form, it is used as a wood binding material. Compounded with a filler, it is used to make housings and small parts.

Unsaturated polyesterresins (UP) (Ungesättigste Polyesterharze) – These resins are colourless and glass clear, with a glossy surface. They can *vary from hard and brittle,* to *soft and elastic*. They have good adhesive strength and good moulding properties.

Epoxyresins (EP) (Epoxidharze) – These resins are colourless to honey yellow in colour, *tough* and *unbreakable*. They have good adhesive, casting and moulding properties. They are used as adhesive resins, resins for paints, casting resins, and in glass-reinforced fabrics. They are also used in the manufacture of *composite materials* like resins reinforced with glass or carbon fiber. These are used to fabricate, boats, aircraft parts, sports equipment and corrugated sheets.

Polyurethane resins (PUR) (Polyurethanharze) – These are honey yellow in colour, transparent and vary from *hard and tough,* to *soft and elastic* . They have good adhesive properties and can be used to produce foam materials. They are *resistant* to *weak acids, alkalies, salt solutions and solvents.* Pure resins are used to make gear wheels, bearing boxes,etc. Medium hard resin is used to make toothed belts, bumpers for cars, rollers, etc. Soft resins are used for gaskets and packings. Polyurethane resins are also used for paints and adhesives.

Silicon resins (SI) (Silikonharze) – These are milk white in colour and vary from *stiff and solid,* to *soft and elastic*. They repel water and adhesives, but show poor resistance to acids, alkalies and solvents. They are used in insulating paints, water repelling paints, gaskets, moulds for castings, etc.

4.4.3 Elastomers (Elastomere)

The most important characteristic of elastomers like natural rubber is their *high degree of elasticity*, meaning their ability to return to their original form after they have been subjected to *large deforming forces*. Elastomers need *treatment before use*, because in the raw state they are soft and sticky when heated, and hard and brittle when cooled.

Natural rubber (Naturkautschuk) — Natural rubber is produced by coagulating the latex of the rubber tree. Freshly cut rubber has the property of self adhesion. For most applications, the rubber is *vulcanized*. In this process, the rubber is made to combine with sulphur or other chemical substances. This process improves mechanical properties, reduces stickiness, makes it insoluble in solvents, and enables it to be less affected by temperature changes. *Carbon black* is added to increase *tensile strength* and *resistance to abrasion*. Various other substances are added like colouring pigments, protective agents, and vulcanizing accessories. Natural rubber is used in a multitude of products like auto tyres, rubber springs, conveyor belts, rubber treads, etc.

Synthetic rubber (or Butadiene – Styrene copolymer (GR-S)) (Synthesekautschuk oder Butadienkautschuk)
GR-S is the most most used synthetic elastomer. The two materials butadiene and styrene (which are petroleum products) are copolymerized to form GR-S. It has better *wear resistance* and *temperature properties* than natural rubber. It is used to produce auto tyres, gaskets, shoe soles, hoses, conveyor belts, etc.

Nitrile rubbers (Acrylnitryl- Butadien Kautschuk)
These are the elastomers that are most used to produce objects that are *resistant to oil, fat and fuel*. These are not resistant to benzol, glycol based brake fluids, etc. They are used to make benzene hoses, gaskets, membranes, etc.

Polyurethane rubbers (Polyurethankautschuk) – These have *twice the tensile strength* of conventional rubber. Solid as well as foamed articles can be made from these elastomers. Used to make guide wheels, shock absorber parts, foam articles, etc.

Silicon rubbers (Silikonkautschuk) – These can be used over a wide range of temperatures (-100°C to +200°C) and are *flame-resistant*. They are resistant to oils and fats, but not to hydrocarbons, fuels, acids and alkalies, or to hot water and steam.

Butyl rubber (Butylkautschuk) – This elastomer has good damping properties and good *ageing properties*. It is resistant to acids, alkalies, acetone and hydraulic fluids but not resistant to fats, fuels and hydrocarbons.It is used to make inner tubes for tyres, insulation, damping elements, hot water hoses, etc.

4.5 Composite materials (Verbundwerkstoffe)

Composite materials are used when a single material does not have the required properties. Examples of commonly used composite materials are *reinforced concrete* and *wood*, which is a *natural composite material* composed of cellulose and lignin. Alloys are not classed as composite materials, but laminated materials together with fibre or particle reinforced materials warrant this description.

4.5.1 Fibre-reinforced materials (Faserverstärkte Werkstoffe)
The reinforcing fibre that is most used is glass fibre. It has *great strength, low density* and is *relatively cheap*. The fibre can be in the form of strands, mats, or in a woven form like cloth. For applications where even greater strength is required as in aircraft construction, *carbon, metal or ceramic fibres* are used.

Glass fibre-reinforced materials consist of thermosetting plastics like polyester or epoxy resins reinforced by glass fibre. They are used to make objects like tennis racquets, gear wheels, automobile bodies, boats, aircraft components, etc.

4.5.2 Particle reinforced materials (Teilchenverstärkte Verbundwerkstoffe)
These are made by using a *thermosetting plastic* reinforced by suitable *fine particle materials*. The thermosetting plastics include melamine, phenolic resins or polyester resins. These are stronger than parts made of the pure plastic. They are used for small parts, electrical components, housings, etc.

Polymer concrete (Polymerbeton) – This is a particle-reinforced material made from epoxy resin and a filler composed of granite particles. Bodies of machine tools made from this material have better *damping properties* than bodies made from grey cast iron. This results in an improvement in the *accuracy* of the parts produced by the machine tools.

Grinding wheels and honing tools (Schleifscheibe und Honwerkzeuge)
These are abrasive grinding stones in which the abrasive material is aluminium oxide, silicon carbide or diamond, bonded in a plastic, ceramic, or metal body. In the grinding stones, the cutting is done by the abrasive particles, while the body acts as a bonding medium giving the tool strength and toughness.

4.5.3 Laminated materials (Schichtverbundwerkstoffe)
In materials like plywood, laminated plastics and paper, thin layers of material are covered with an adhesive, placed one over another and then pressed together. The *grain in each layer* (as for example the wood grain in plywood) is *at right angles* to the adjoining layer, giving the final laminated product *good strength in all directions*.

4.6 Lubricants (Schmierstoffe)
Most machines have surfaces and bearings in which two surfaces are in contact and move relative to each other. It is necessary to *reduce the friction* between the surfaces so that the *wear, damage* and *heat generated* can be *minimized*. This is done by a process called lubrication in which a *friction reducing substance* is introduced between the surfaces. The substance may be a solid, liquid or a gas. Lubricants are manufactured to have specific physical and chemical properties. An appropriate lubricant is chosen to suit the particular application. Chemicals called *additives* are added to modify certain

characteristics of a lubricant. *Viscosity* is probably the most important property of a lubricant and is a function of the temperature, pressure and flow of the lubricant.

4.6.1 Liquid lubricants (Flüssige Schmierstoffe)

Many liquids including water are used as lubricants, but the ones most used are of two types (a) lubricants obtained from refined petroleum
(b) synthetically produced lubricants.

4.6.2 Lubricants from petroleum (Mineralöle aus Erdöl)

These lubricants are oils obtained from petroleum. *Additives* are added to these lubricants to improve their resistance to corrosion, resistance to the action of high pressures, and to improve their ageing properties. Although the *viscosity* of these oils *change with temperature*, satisfactory performance over a wide range of temperatures is possible by using additives.

4.6.3 Synthetic oils (Synthetische Öle)

These oils are *synthesized from hydrocarbons*. Synthetic oils have a better viscosity versus temperature relationship than petroleum lubricants, and better ageing properties, but are relatively *expensive*.

4.6.4 Lubricating greases (Schmierfette)

Lubricating greases are made by adding a *soap-like substance* to mineral or synthetic oil. This converts it into a paste-like substance called *grease.* Other substances like clay, chemicals and polymers may also be added. Greases are used to lubricate *ball* and *roller bearings* and heavily loaded *guideways*.

4.6.5 Solid lubricants (Festschmierstoffe)

Solid lubricants are used when the *relative speed* between two sliding surfaces is *too small* for the building of an *oil film*, and also under severe conditions of temperature and pressure.

The simplest are *unbonded lubricants* in powdered form. The most commonly used substances are graphite, molybdenum disulphide, PTFE and also metal oxides, talc and salts. The life of these lubricants may be limited and *longer life* may be realized by using *bonded solid lubricants*. The solid lubricants are mixed with binding materials (binders) and applied to the sliding surfaces.

4.6.6 Gases (Gase)

Gases have low viscosity coefficients. In order that the two sliding surfaces may be separated, small holes are drilled in the bearings and the gas is fed into these under pressure.

4.7 Additives and fillers (Zusatzstoffe und Füllstoffe)

It is often necessary to introduce *additional substances* (called additives) into materials in order to *modify* and *improve their properties*. The added substances fall into two groups. Substances which enter the molecular structure of the materials are called additives, while those that remain separate are called fillers.

4.7.1 Additives (Zusatzstoffe)

Additives are usually added in small quantities, but usually have a *marked effect* on the *properties* of a material. There are many types of additives which can perform different tasks. These can be added to plastics, paints, lubricants, fuels and to a wide range of mixtures and compounds.

Some types of additives used are listed below.

1. **Accelerator** (Beschleuniger)
2. **Antioxidants** (Oxydierungsschutzmittel)
3. **Dyes and pigments** (Farbstoffe und Pigmente)
4. **Lubricants and flow promoters** (Schmiermittel und Stromförderungsmittel)
5. **Plasticizers** (Plastifizierungsmittel)
6. **Solvents** (Lösungsmittel)

4.7.2 Fillers (Füllstoffe)

Fillers are used to reduce material costs, to ease processing, reduce shrinkage, and to increase the electrical or thermal conductivity. They are also used as *reinforcing materials*. Some of the *fillers* used are listed below.

1. **Alumina**	Tonerde	5. **Talc**	Talk
2. **Clay**	Lehm	6. **Wood flour**	Holzmehl
3. **Minerals**	Mineralien	7. **Synthetic**	Synthetische
4. **Quartz**	Quartz.	**fibres**	Fasern

5 The testing of materials (Werkstoffprüfung)

5.1 Tensile tests (Zugversuch)

This test has already been described under **II Strength of materials, section 1.6.2** (p 44). As mentioned there, quantities like *yield strength*, *ultimate tensile strength* (UTS) and *breaking strength* of a material need to be known, before a material can be selected for a particular task.

5.2 Notched bar (Izod) impact test (Kerbschlagbiegeversuch)

In this test, a metal test piece is subjected to a violent blow given by a heavy pendulum. If the pendulum makes an angle ϕ_0 with the vertical when released, it would make almost the same angle when it swings to the other side if no test piece was present. The test piece is accurately made to the dimensions specified in the test and usually has a *notch* cut in it (Fig 3.4). If test piece is now placed in the path of the pendulum, it will either be bent or broken on impact. The pendulum will *lose energy* and its swing on the other side will be *reduced.*

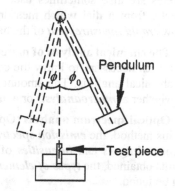

Fig 3.3 Principle of the impact test

The value of the angles ϕ_0 and ϕ can be read off on a scale (not shown in the figure). The energy loss of the pendulum corresponding to each specimen can be calculated from the difference in the *maximum heights* of the pendulum before and after the impact. The smaller the angle ϕ, the tougher the material. One possible pendulum arrangement is shown in Fig 3.4.

5.3 Hardness tests (Härteprüfungen)

Hardness has been defined as the ability of a material to withstand *abrasion* or *indentation*. In components and tools, quite often only the surface is hardened, while the interior remains tough. *Minerals* are classified for hardness on the *Mohs scale* for minerals. In this scale ten minerals are so arranged that a pointed fragment of each mineral will scratch the next mineral lower down on the scale.

In industry, the resistance to *indentation* or *penetration* is widely used as a test for the hardness of metals and other engineering materials. Some of the indentation tests used are briefly described below.

5.3.1 Brinell hardness test (Härteprüfung nach Brinell)

In this test a hardened steel sphere is forced under a known load into the surface of a material and the *diameter of the indentation* created is measured. The Brinell hardness number is obtained by dividing the load in kg by the surface area of the indentation.

5.3.2 The Vickers hardness test (Härteprüfung nach Vickers)

This test uses an indenter with a *diamond tip* having the form of a pyramid. Since diamond is the hardest substance available, this test can be used *to test hardened cutting tools*. Tables are usually provided with each instrument giving the hardness number in terms of the *length of the diagonals* of the indentation.

5.3.3 The Rockwell hardness test (Härteprüfung nach Rockwell)

This differs from the above tests in that the hardness is determined by the *depth of penetration* of an indenter with a radiused diamond tip. Hardened steel spheres are also sometimes used. The reading for hardness can be obtained directly from a dial which measures the penetration. This type of instrument allows *rapid measurement* of the hardness and is much used in *production*.

5.4 The chemical analysis of materials (Chemische Prüfungen)

It is often necessary to know the composition of materials and this can be done by chemical analysis. The amount of material needed for analysis will depend on whether a *macroanalysis* or a *microanalysis* is planned.

5.5 Optical spectrum analysis (Optische Spektralanalyse)

In this method, the *emission spectrum* of the material is first produced and both the *wavelengths* and *intensities* of the spectrum lines are measured. From the results obtained, the *types of elements* in the sample and their *relative quantities* can be found.

5.6 Fatigue tests (Dauerfestigkeitsprüfung)
Materials which are subjected to *repeated loading* at levels well *below the yield strength* of the material, can gradually deteriorate and fracture. To avoid this kind of failure, fatigue tests are often made on materials. In these tests, the specimen is subjected to *periodically varying stresses* of constant amplitude. At higher stress values, the material fails after a number of cycles. By lowering the value of the applied stress, a value can be found which *does not produce failure* regardless of the number of cycles.

5.7 X´ray fluorescence analysis (Röntgenfluoreszenzanalyse)
Here again as in the optical spectrum, the X´ray spectrum of the sample is produced, and from a study of the spectrum, the components may be identified.

5.8 Electron beam microanalysis (Elektronenstrahlanalyse)
In this method, the *composition of a tiny area* of a material can be studied. A very fine electron beam strikes the sample which emits X´rays which are characteristic of the elements in the small area of the sample.

5.9 Metallographic analysis (Metallographische Untersuchungen)
Both macroscopic and microscopic metallographic studies are possible. *Macroscopic studies* are able to reveal the existence of cracks, pores, fractures, etc. *Microscopic studies* are carried out on etched and polished samples. The *microstructure* of the *crystal grains* is *visible under the microscope,* and a study of their composition, shape and orientation, gives much information about the state of the specimen.

5.10 X´ray and γ´ray tests (Röntgen- und Gammastrahlenprüfungen)
X´rays can be used to detect *internal defects* like cracks, porosity, inclusions, corrosion, etc. in metallic, nonmetallic and composite materials. Gamma rays may be used in cases where it is difficult to use X´ray equipment as for example where *electrical power* is not available or where the source cannot be placed in a *particular position*. The methods are similar to those for X´rays.

5.11 Magnetic particle tests (Magnetische Rißprüfungen)
This method can be used to detect *surface defects* in *ferromagnetic materials.* The object to be tested is magnetized and then finely divided magnetic particles are sprinkled on its surface. Any field discontinuity due to defects or cracks attracts the particles and gives a *visible indication* of the discontinuity.

5.12 Eddy current tests (Wirbelstromprüfungen)
Eddy currents are induced when a metal is placed in a varying magnetic field. The eddy currents create a magnetic field which *opposes* the *inducing magnetic field*. Imperfections and discontinuities in the metal cause a *change in* the *apparent impedance* of the *field producing coil* or the detector coil. This method can be used to investigate cracks, inhomogeneities, thickness, case depth, composition, hardness, heat treatment, etc. Care has to be taken in interpreting the results. The frequencies used lie between 1 Hz and 5 MHz.

4 Thermodynamics (Thermodynamik)

1 Basic concepts and temperature(Grundkonzepte und Temperatur)

1.1 Macroscopic and the microscopic points of view
(Makroskopische und Mikroskopische Betrachtungsweise)

In any type of scientific study, attention is focused on a *region of space* or a *finite portion of matter* which we can call a *system*. Anything which is outside the system and which affects its behaviour can be called the *surroundings*.

The behaviour of a system can be studied from two separate points of view, *macroscopic* and *microscopic*. In studying the behaviour of a system, we must choose *suitable quantities* (or *coordinates*) which define the *states of a system*.

A macroscopic study of a system uses a few *large scale properties* which are *directly measurable*. No assumptions concerning the *structure of matter* are made. For example if we need to study the behaviour of a gas, useful coordinates would be composition, volume, temperature and pressure.

A microscopic study of a system involves the *small scale properties* of a system. *Assumptions* have to be made regarding the *structure of matter* in such a system. For example a gas can be supposed to be composed of a *large number* of *molecules* moving at *high speeds* in a container. In this case a *large number of quantities* have to be specified, which are *not directly measurable*. Only by using *statistical methods* is a satisfactory study of such a system possible.

Although these *two points of view* are *different*, both deal with the *same system* and arrive at the *same conclusions*. Hence there must be a *relationship* between them. This lies in the fact that the macroscopic description of the system uses as coordinates a *few directly measurable quantities,* which are the *time averages* of a large number of microscopic quantities.

1.2 Thermodynamic systems and the state of a system
(Thermodynamische Systeme und der Zustand eines Systems)

Thermodynamics uses the *macroscopic point of view* to study *particular types* of systems and focuses on the *interior of the systems.* A system can be in one of seve.al different states, and each system requires a definite number of *macroscopic quantities* or *coordinates* to *define* each of its states. For example *each state* of an *ideal gas* can be defined by two coordinates P and V.

The *type* and *number* of coordinates required for the description of each *state* of a given *thermodynamic system* are found by *experiment*. Such coordinates are called *thermodynamic coordinates*.

1.3 Thermal equilibrium and temperature
(Thermisches Gleichgewicht und Temperatur)

It has been seen that the state of a system may be specified in terms of a few macroscopic coordinates. A system is said to be in a *state of equilibrium* if its *coordinates do not change* as long as the *external conditions* do not change.

If we have two systems close to each other *separated by a wall*, the thermal influence exerted by one system on the other, depends on the *type of wall* that separates them. Walls can be of two types, *adiabatic* or *diathermic*. An adiabatic wall is a wall that forms a *thermal barrier* between the systems. A diathermal wall (like a thin metal sheet) allows a free *thermal flow* between the systems until the two systems are in *equilibrium with each other*. The two systems become part of a *combined system* which is in *thermal equilibrium*.

Zeroth law - If *two systems* are in thermal equilibrium with a *third system*, they are in thermal equilibrium with *each other*. This statement is called *the zeroth law of thermodynamics.*

1.3.1 The concept of temperature (Das Temperaturkonzept)

Temperature has been said to be a measure of the *degree of hotness* in a body. Consider two systems A and B, each of which require only *two independent coordinates* for the *specification* of a *state of the system*. *System A* which is in a state having coordinates (X_1, Y_1) is in *thermal equilibrium* with a *system B* which is in the state (X'_1, Y'_1). If the system A is removed away from B and changed,

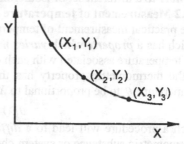

Fig 4.1 An isotherm of system A

it is found that *there are a number of states* (X_1, Y_1), (X_2, Y_2), (X_3, Y_3), ... each of which is in *equilibrium* with the *same state* (X'_1, Y'_1) of system B, and all of which by the *zeroth law of thermodynamics* are in equilibrium with each other. When all these points are plotted on a diagram (Fig 4.1), we have a curve which can be called an *isotherm*. An isotherm can be stated to be the locus of all points corresponding to the states of a system which are in *thermal equilibrium* with *one state* of another system.

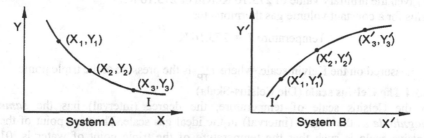

Fig 4.2 Corresponding isotherms for two systems

Similarly with system B, we can find we can find different states (X'_1, Y'_1), (X'_2, Y'_2), (X'_3, Y'_3), ... which are in equilibrium with one state (X_1, Y_1) of system A, and therefore in thermal equilibrium with one another. If these states are plotted on a diagram, we can obtain an isotherm I′. It follows that all states on the isotherm I of system A are in equilibrium with all states on isotherm I′of system B. The curves I and I′ may be called *corresponding isotherms*. All states belonging to corresponding isotherms of all systems have something in common, this being that they are all in *thermal equilibrium* with *each other*.

This property which determines whether a system is in thermal equilibrium with other systems can be called *temperature*. Temperature is a *scalar quantity* and the temperature of all systems in thermal equilibrium with each other may be represented by a *number*. A *temperature scale* is established by adopting a *set of rules* to *assign a number* to a set of corresponding isotherms and a *different number* to a different set of isotherms.

1.3.2 Measurement of temperature (Messung der Temperatur)

The practical measurement of temperature is accomplished by selecting a system which has a *property* which *varies with temperature* and assigning a number to the temperature associated with each of its isotherms.

If the thermometric property has the coordinate X, then we can assume the temperature θ to be proportional to X and write,

$$\theta(X) = aX$$

Such a procedure will lead to a *different temperature scale* for each different thermometric substance or system chosen. In the end only one temperature scale will have to be selected as the *most fundamental* and used as a *standard temperature scale*.

1.3.3 Fixed points on a temperature scale (Fixpunkte einer Temperaturskala)

To have a temperature scale, it is necessary to have one or more *arbitrarily fixed points* on the scale. In the past, two fixed points had been used, the melting point of ice and the boiling point of water. More recently only one point has been used, the *triple point* of *water* which is the point at which pure water exists as an *equilibrium mixture* of *ice, liquid* and *vapour*. The temperature of this point is given the arbitrary value of 273.16 Kelvin or 273.16 K.

Thus for a constant volume gas thermometer

$$\text{Temperature } T = 273.16 \text{ K } \frac{P}{P_{TP}}$$

as measured on the Kelvin scale, where P_{TP} is the pressure at the triple point.

1.3.4 The Celsius scale (Die Celsius-Skala)

In the Celsius scale of temperature, the degree (interval) has the *same magnitude* as the degree (interval) in the ideal gas scale. The zero point of the Celsius scale is such that the temperature of the triple point of water is .01 degree Celsius (or 0.01°C). Thus if θ is the temperature on the Celsius scale and T the temperature on the Kelvin (or absolute) scale, then

$$\theta \,°C = T(K) - 273.15$$

1.3.5 The Kelvin (or absolute) scale and the ideal gas scale
(Die Kelvin-(oder absolute) Skala und die ideale Gas-Skala)

A practical scale of temperature like the ideal gas scale is dependent on the *properties* of the *thermometric substance* used. It can however be shown that there exists a *scale* of temperature which is *independent of the properties* of the substance used. Such a scale of temperature is called an *absolute scale* of temperature and the temperature on this scale is measured in degrees Kelvin.

An absolute scale of temperature would remain an *abstraction* unless a way of realizing such a *scale in practice* can be found. It can be shown that the *ideal gas scale* is *numerically identical* to the absolute or Kelvin scale of temperature, and that the temperatures measured on the *ideal gas scale* are the *same as those* on the *absolute scale* in degrees Kelvin. For this reason, the ideal gas scale is used as the *fundamental scale* for the measurement of temperature.

1.4 Thermal expansion (Wärmeausdehnung)
When a body is heated, its dimensions change.

The *coefficient of linear expansion* (or the *linear expansivity*) of a solid is defined as the increase in length per degree rise in temperature divided by the length at 0°C.

The *coefficient of areal expansion* of a solid (or *areal expansivity*) is defined as the increase in area per degree rise in temperature divided by the area at 0°C.

The *coefficient of volume expansion* of a body (or the *volume expansivity*) is defined as the increase in volume per degree rise in temperature divided by the volume at 0°C.

We can write

Coefficient of linear expansion $\quad \alpha = \dfrac{l_t - l_0}{l_0 t} \quad$ and $\quad l_t = l_0(1 + \alpha t)$

Coefficient of areal expansion $\quad \beta = \dfrac{A_t - A_0}{A_0 t} \quad$ and $\quad A_t = A_0(1 + \alpha t)$

Coefficient of volume expansion $\quad \gamma = \dfrac{V_t - V_0}{V_0 t} \quad$ and $\quad V_t = V_0(1 + \alpha t)$

For a homogeneous isotropic solid, it can be shown that $\beta = 2\alpha$ and $\gamma = 3\alpha$

The coefficients of expansion for solids are small (much smaller than for liquids and gases), However large lengths of a solid expand to such an extent, that gaps have to be left to allow for expansion. A good example is the gap that is left between two lengths of steel rails in a railway line, or in the supporting beams of a bridge. Failure to allow for expansion can cause serious damage.

2 Thermodynamic systems and work
(Thermodynamische Systeme und Arbeit)

2.1 Thermodynamic equilibrium (Thermodynamisches Gleichgewicht)

The number of thermodynamic coordinates which are *necessary* and *sufficient* to provide a macroscopic description of a particular system can be *found by experiment*. The *state* of a *thermodynamic system* is determined by the *values* of its *thermodynamic coordinates.*

(It is necessary to state here that that the expression *change of state* does *not refer* to a transition from solid to liquid, or liquid to vapour. In thermodynamics such changes are referred to as *changes of phase*.)

For a system to be in a state of *thermodynamic equilibrium*, it must satisfy the conditions for three other types of equilibrium, *mechanical, thermal* and *chemical.*

 a) A system is in a state of *mechanical equilibrium* when there is no *unbalanced force* in the interior of the system and also between the system and its surroundings.

 b) A system which is in mechanical equilibrium is also in *chemical equilibrium*, when it does not experience a change in its *internal structure*, such as a *chemical reaction* or a *transfer of matter* from one part to another, by processes such as *diffusion* or *solution*.

 c) If a system is in mechanical and chemical equilibrium, then it is also in *thermal equilibrium* when it is separated from its surroundings by a *diathermic wall*. In thermal equilibrium, *all parts* of a system are at the *same temperature*, and the temperature is the same as that of its *surroundings*. When this is not the case, a *change of state* will take place until thermal equilibrium is achieved.

When *all three conditions* are satisfied, the system is in a state of *thermodynamic equilibrium*. Under these conditions, there is no tendency for any *change of state* by the system or its surroundings. States corresponding to thermodynamic equilibrium can be described in terms of *thermodynamic coordinates*. These are *macroscopic coordinates* which *do not involve the time*.

2.2 Equations of state (Zustandsgleichungen)

The *concept* of an *equation of state* can be conveniently illustrated by considering the behaviour of a *constant mass* of an *ideal gas* whose pressure P, volume V and temperature T can be measured. If the magnitudes of P and V are fixed, then we know that T also assumes a *fixed value*. This means that of the three thermodynamic coordinates, *only two* are *independent variables* and that there *exists a relationship* between the three coordinates which *deprives one* of them of their *independence*.

An equation of state is a *relationship* between the *thermodynamic coordinates* of a system which is in thermodynamic equilibrium. Different systems have *different equations* of *state,* and an equation of state has to be determined *experimentally* for each system. It cannot be determined by thermodynamics alone. It is an *experimental addition* to thermodynamics.

2.3 Hydrostatic systems (Hydrostatische Systeme)

An isotropic system which has a *constant mass* and exerts a *uniform hydrostatic pressure* on its surroundings is called a hydrostatic system. Pure substances, homogeneous and heterogeneous mixtures of solids, liquids and gases are examples of hydrostatic systems. Experiments have shown that the states of equilibrium of a hydrostatic system can be described in terms of *three macroscopic coordinates*. These are, the pressure P which the system exerts on its surroundings, the volume V and the absolute temperature T.

2.4 Work (Arbeit)

If a force acts on a system and the system *undergoes a displacement* under the action of this force, then work is said to be done. The amount of work done is equal to the product of the *magnitude of the force* and the *component* of the *displacement* in the direction of the force.

When a system as a whole exerts a *net force* on *its surroundings* and a displacement takes place, the work done is called *external work*. This external work can be done *by the system* or *on the system*. The term *internal work* refers to work done by *one part* of a system *on another*.

Only external work which involves an *interaction* between a *system* and its *surroundings* is of significance in thermodynamics. Internal work is *not considered* in this context, and the term work as used in thermodynamics means *external work*.

2.5 Quasi-static processes (Quasi-statische Prozesse)

When a system is in a state of thermodynamic equilibrium, there is no tendency for a change of state in the system or its surroundings. A *change of state* of the system is only possible if a *finite unbalanced force* acts on the system. The action of a finite unbalanced force however *upsets the conditions* for mechanical and consequently *thermodynamic equilibrium*. A finite unbalanced force causes turbulence, acceleration, etc. and makes the system to go through *nonequilibrium states*.

When a system undergoes a change, it goes through a *succession of states*. If the successive states of a system in a process of change are to be described by thermodynamic coordinates, then it is clear that the change *must not be caused* by a *finite unbalanced force*. We can however think of a situation, where the change is caused by an *unbalanced force* which is *infinitely small*.

Such a process is called a *quasi-static process* and in such a process the system is always *infinitesimally close* to a state of *thermodynamic equilibrium*. In such an *ideal process*, all the states through which the system passes can be described by *thermodynamic coordinates*. It follows that an equation of state holds for all these states.

2.6 Work done by a gas (Arbeit die ein Gas verrichtet)

One way of effecting the *transfer of energy* between a system and its surroundings is by doing work. Work *done by a system* on its surroundings is considered to be *negative*, while work *done on the system* by the surroundings is considered to be *positive*.

Consider a gas of volume V contained in a cylinder, at a pressure P and at a temperature T. This system is in a state of *thermodynamic equilibrium*. The pressure of the gas P is *balanced* by the pressure P exerted by the surroundings. If the volume of the gas *increases* by a small amount dV and the process of change is a *quasi-static process*, the work done *by the gas* is given by

$$dW = -PdV$$

If the volume of the gas is now changed in a quasi-static process from an initial volume V_i to a final volume V_f, the work done by the gas is

$$W_{if} = - \int_{V_i}^{V_f} PdV$$

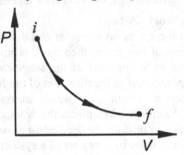

If we move along *the same path* in the *opposite direction*, the work done on the gas is

Fig 4.3 Expansion and contraction along the same path

$$W_{fi} = - \int_{V_f}^{V_i} PdV \ .$$ Since the process is quasi-static, we can write

$$W_{if} = -W_{fi}$$

2.7. A cyclic process (or cycle) (Ein Kreisprozess)

A series of processes represented by a *closed figure* is called a *cycle* and the enclosed area represents the *net work* done by a system. A cycle for a gas consisting of *two stages*, an *expansion* and a *compression* is shown in Fig 4.4.

The changes in P and V during the *expansion* are shown in Fig 4.4 (a). The work done is *negative*, and is given by the shaded area under the curve. Fig 4.4 (b) shows the work done during the *compression*. Here the work done is *positive*. In Fig 4.3 (c) both curves are drawn together to show that after the two processes, the gas is in its *initial state*. The enclosed shaded area represents the *net work* done. Here the *direction* of *traverse* of the cycle is such that net work is *done by the gas,* and this is *negative*. If the direction of traverse was *reversed*, the net work is *done on the gas* and this would be *positive*.

It is clear that the work done *depends on the path*. This means that the work done depends *not only* on the *initial* and *final states,* but also on the *intermediate states*.

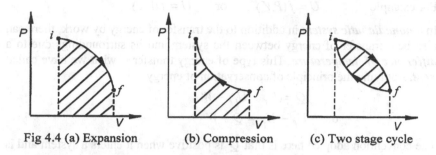

Fig 4.4 (a) Expansion   (b) Compression   (c) Two stage cycle

3 The first law of thermodynamics (Erster Hauptsatz)

3.1 Heat and work (Wärme und Arbeit)

Thermodynamics concentrates on the study of *two quantities, heat* and *work*. We know that heat is a quantity that flows from a body at a higher temperature to a body at a lower temperature. Heat can therefore be conveniently defined as a quantity that can be *transferred* between a *system* and its *surroundings*. An *adiabatic wall* is one that *prevents* the flow of heat, and is therefore a heat *insulator*. A *diathermic wall* is one that *allows* a free flow of heat. An example of a diathermic wall is a sheet of metal.

A system that is completely insulated from its surroundings by an adiabatic envelope does not allow the transfer of heat between the system and its surroundings. However it can be *mechanically coupled* to its *surroundings* so that *external work* can be done. Such work is known as *adiabatic work* and the state of a system can change from a given initial state to a final state by the performance of adiabatic work only. Experiments show that the amount of adiabatic work *does not depend* on the *path* (or on the *intermediate states*), but only on the initial and final states of the system.

3.1.1 Internal energy function (Innere Energiefunktion)

From this it follows that for a thermodynamic system, there exists a function whose *final value* minus its *initial value* is equal to the adiabatic work done in moving from the initial state to the final state. This function is known as the *internal energy function* and is denoted by U. We can therefore write

$$\text{Adiabatic work} \quad W_{if} = U_f - U_i$$

The difference $U_f - U_i$ represents the increase in the *internal energy* of the system. We have seen that the equilibrium states of a hydrostatic system can be described in terms of three coordinates P, V, T.

If an *equation of state exists,* then only two coordinates are required. We can therefore represent the internal energy as a function of *two thermodynamic coordinates*.

For example $\qquad U = f(P,V) \qquad$ or $\qquad U = f(V,T)$

In a *nonadiabatic system* in addition to the transfer of energy by work, there can also be a transfer of energy between the system and its surroundings due to a *difference of temperature*. This type of energy transfer is what we have called *heat*. Applying the principle of conservation of energy

$$Q = (U_f - U_i) - W$$
$$\text{or} \quad U_f - U_i = Q + W$$

The convention adopted here is that Q is positive when it enters a system and is negative when it leaves the system.

The above relationship is known as the *first law of thermodynamics*.

We have seen that the work done on or by a system depends on the path by which the system is moved from the initial to the final state. It follows that *both heat and work* are *not functions* of *thermodynamic coordinates* and are *inexact differentials*.

For an infinitesimal quasi-static process, we can write the first law in the form

$$dU = dQ + dW$$

For an infinitesimal quasi-static process of a *hydrostatic system*, this becomes

$$dU = dQ - PdV$$
$$\text{or} \qquad dQ = dU + PdV$$

3.2 Heat capacity (Wärmekapazität)

When a quantity of heat is absorbed by a system and this results in a change of temperature from T_i to T_f , the average heat capacity of the system is defined as

$$\text{Average heat capacity} = \frac{Q}{T_f - T_i}$$

For an infinitesimal change of temperature dT this may be written as

$$C = \frac{dQ}{dT}$$

The *specific heat capacity* of a substance c (usually abbreviated to *specific heat*) is the amount of heat required to increase the temperature of 1 kg of the substance by 1 K without any change in phase.

The units of c are J/kg K or J/kg °C.

An important quantity is the ***molar heat capacity***. This is the heat capacity corresponding to the molar mass M of a substance. The molar heat capacity is measured in J/mol. K or J/mol.°C. The specific heat (per kg) and the molar heat capacity are designated by c, while the heat capacity for any arbitrary mass is designated by C.

3.3 Specific heat capacities c_P and c_V (Spezifische Wärmekapazitäten c_P & c_V)

Two important values for the specific heat are the specific heat at constant pressure c_P and the specific heat at constant volume c_V. They are given by the expressions

$$c_P = \left(\frac{dQ}{dT}\right)_P$$

$$c_V = \left(\frac{dQ}{dT}\right)_V$$

c_P and c_V usually have different values.

For ideal gases it can

be shown that $\qquad c_P - c_V = R_i$

    and that $\qquad c_P / c_V = \kappa$

4 The second law of thermodynamics (Zweiter Hauptsatz)

4.1 Conversion of heat into work and work into heat
(Umwandlung von Wärme in Arbeit und Arbeit in Wärme)

It is a matter of common experience that ***work*** can be ***completely transformed*** into ***heat***. For example when we rub two objects together, the work done against frictional forces is transformed into ***internal energy*** causing a rise in temperature. In the converse process where heat is converted into work, a ***100 %*** conversion is not possible. Only a partial conversion is possible and this is

usually done in a heat engine. The principle involved in a heat engine is shown in Fig 4.5. A heat engine works between two heat reservoirs. It absorbs heat from the hot reservoir, converts some of this into work, and rejects the remainder into the cold reservoir. This contains the basic concept involved in the ***second law of thermodynamics***. This law has been stated in many ways. One statement is the following:

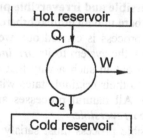

Fig 4.5 Principle of the heat engine

Second law of thermodynamics (first statement): No process has been developed which converts the heat extracted from one reservoir into work without rejecting some heat into a reservoir at a lower temperature.

4.2 The heat engine (Die Wärmekraftmaschine)

In a heat engine an amount of heat Q_1 is absorbed from a hot reservoir and an amount of heat Q_2 is rejected into a cold reservoir. The quantity Q_1 is larger than Q_2 and part of the difference $Q_1 - Q_2$ is converted into work W.

The thermal efficiency $\eta = \dfrac{\text{work output}}{\text{heat input}}$

$$\eta = \frac{W}{Q_1}$$

From the first law $W = Q_1 - Q_2$

Therefore $\eta = \dfrac{Q_1 - Q_2}{Q_1}$

4.3 The refrigerator (Der Kühlschrank)

The refrigerator is a device which works in an **opposite manner** to the heat engine. It absorbs some heat at a low temperature and rejects a larger amount of heat at a higher temperature. Work has to be done on the refrigerator for this to take place. The principle involved is shown in Fig 4.6. Another statement of the second law of thermodynamics **based on the refrigerator** is possible and this is:

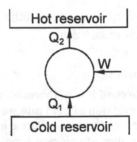

Fig 4.6. Principle of the refrigerator

Second law of thermodynamics (second statement):   It is not possible to have a system which transfers heat from a cold body to a hot body without work being done on the system.

4.4 Reversible and irreversible processes

(Umkehrbare und nichtumkehrbare Prozesse)

When a process is carried out, we can in many cases **reverse the process** and bring back the system to its **original state**. A truly reversible process is one that is carried out in such a way that both the system and its surroundings can be restored to their original states without **causing any changes** in the **rest of the universe**. All natural processes are **unable** to fulfil these conditions and are therefore **irreversible.**

A reversible process must satisfy the conditions for mechanical, chemical and thermal equilibrium. The process itself must be **quasi-static** which means that it must pass through a series of **thermodynamic equilibrium states**. In addition, **no dissipative effects** such as friction, viscosity, magnetic hysterisis, etc, should be present. Since natural processes do not satisfy these conditions, it follows that a reversible process is *an abstraction,* which is nevertheless very *useful for theoretical purposes.*

4.5 Entropy (Entropie)

It has been shown by Clausius that for any

reversible cycle $\oint_R \dfrac{dQ}{T} = 0$

Let us consider a reversible cycle in which a system first moves from an initial state i to a final state f along a reversible path R_1. It then returns along a reversible path R_2 from f to its initial state i. For this cycle, we can write

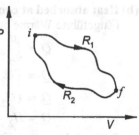

Fig 4.7 Two reversible paths

$$\oint_R \frac{dQ}{T} = \int_i^f \frac{dQ}{T} \text{ along } R_1 + \int_f^i \frac{dQ}{T} \text{ along } R_2 = 0$$

$$\int_i^f \frac{dQ}{T} \text{ along } R_1 = \int_i^f \frac{dQ}{T} \text{ along } R_2$$

This shows that the integral from i to f *does not depend on the path*, but only on the *initial* and *final states*. It follows that a function S exists whose change

is given by $\qquad \Delta S = S_f - S_i = \int_i^f \dfrac{dQ}{T}$ along a reversible path.

This function S has been given the name entropy. The above integral defines only a *change in entropy*, and does not define a value for the *absolute entropy*.

4.6 Changes in entropy (Entropieänderungen)

Changes in entropy of the universe have assumed an important place in scientific thinking. When a system goes through a process, the change in entropy of the system can be added to the change in entropy of the surroundings. This total change can be called the *change in entropy of the universe* due to the process. When a *reversible process* takes place, the entropy of the universe *remains unchanged*. When an *irreversible process* takes place, the entropy of the universe *increases*. Changes in entropy of the universe due to any kind of process (meaning both reversible and irreversible) can be represented as follows:

$$\sum \Delta S \geq 0$$

4.7 Enthalpy (Enthalpie)

The first law of thermodynamics can be used to find the heat absorbed at constant volume and constant pressure for a hydrostatic system.

(a) Heat absorbed at constant volume
(Zugeführte Wärme bei konstantem Gasvolumen)
Using the first law of thermodynamics
$$dQ = dU + PdV$$

If V is constant, $dV = 0$ and $\int_{U_2}^{U_1} dU = U_2 - U_1$

Therefore Q = Increase in internal energy

(b) Heat absorbed at constant pressure
(Zugeführte Wärme bei konstantem Gasdruck)

$$Q = \int_{U_2}^{U_1} dU + P \int_{V_1}^{V_2} dV$$

$$Q = (U_2 - U_1) + P(V_2 - V_1)$$

$$Q = (U_2 + PV_2) - (U_1 + PV_1)$$

$$Q = H_2 - H_1 \quad \text{where} \quad H = U + PV$$

In general we can write

$$H_1 = U_1 + P_1V_1 = H_2 = U_2 + P_2V_2 \quad \text{and} \quad H = U + PV$$

The function H is called *enthalpy*. This function is of special importance because many *thermal processes* take place at *constant pressure*, particularly at *atmospheric pressure*. Phase transitions like boiling, melting, sublimation take place at constant pressure and the latent heat measured is equal to the change in enthalpy. For an infinitesimal change we can write

$$dH = dU + PdV + VdP$$

Since $dQ = dU + PdV$

$$dH = dQ + VdP$$

For a *constant pressure (or isobaric) process*, $dP = 0$ and therefore

$$\left(\frac{\partial H}{\partial T}\right)_P = \left(\frac{\partial Q}{\partial T}\right)_P = c_P$$

Integrating, we have

$$H_f - H_i = \int_i^f c_P\, dT = Q$$

4.8 Phase transitions (Änderung der Aggregatzustände)
Matter normally exists in one of three phases, *solid, liquid* or *vapour*. When a phase transition occurs, the *temperature and pressure* remain *constant*, while the *volume and entropy* change. Phase transitions are usually *reversible*, and one distinguishes between three types of *first order transitions*.

- **Fusion** (or melting) denotes a change from a solid to a liquid phase. The reverse process (which is the change from a liquid to a solid) is called solidification or freezing.

- **Vapourisation** denotes a change from a liquid to a gaseous phase. The reverse process is called condensation. The term *boiling* refers to the phase change which occurs when water changes into steam at normal atmospheric pressure.

- **Sublimation** refers to a direct change from a solid phase to a vapour phase without going through an intermediate liquid phase.

Clear phase transitions between solid, liquid and vapour phases do not always take place. In the case of mixtures and amorphous substances, intermediate stages in a phase transition can occur. The phase in which a substance finds itself depends on the *cohesive* (binding) *forces* present in the substance, and also on the temperature and the pressure. When a solid is heated the atoms and molecules are set into *vibration*. Further heating of the solid causes the *bonds* between the atoms and molecules to be *destroyed*, and allows them to move around in a container almost freely. The solid has now melted and is in a *liquid phase*. Further supply of heat causes even larger movements in the atoms and molecules until a point is reached where the particles are able to overcome all forces between them and *escape* into *free space*. This corresponds to *vapourisation* and the substance is now in a *vapour phase*.

4.8.1 Melting and freezing (Schmelzen und Erstarren)

A substance melts when its phase changes from solid to liquid. In the case of pure substances, there is usually a *definite temperature* at which this occurs called the *melting point*. The temperature *remains constant* during the melting process. Impure substances melt within a range and not at a definite temperature.

4.8.2 Latent heat (or specific latent heat) of fusion (Schmelzwärme)

This is the term used for the amount of energy required to melt 1 kg of a substance at the usual melting temperature. When a substance *solidifies*, the same amount of heat is *released* from the substance.

There is usually an *increase* in the *volume* of a material when melting takes place. The main exception is *water* which undergoes a *decrease in volume* when its melts. The melting point of a solid, whose volume increases when it melts, increases with increasing pressure. The reverse is true for substances like water.

4.8.3 Vapourisation and condensation (Sieden und Kondensation)

The temperature at which the phase transition of a substance from the liquid to the vapour phase takes place is very sensitive to the atmospheric (or other external) pressure. Vapourisation of a substance takes place at a definite temperature, provided the pressure is fixed. This is usually fixed at the standard atmospheric pressure of 1.013 bar. The temperature in °C at which a substance vapourises at a pressure of 1.01 bar is called the *boiling point* of the substance.

4.8.4 Latent heat of vapourisation (Verdampfungswärme) is the term used for

the amount of heat required to change 1 kg of a substance from the liquid phase to the vapour phase at the boiling point of the liquid. The reverse process is called *condensation* or *liquefaction*. Vapourisation involves an enormous increase in the volume of the substance. At lower temperatures vapourisation takes place only from the surface of the liquid and is called *evaporation*. However the latent heat has to be supplied for evaporation to take place.

5 Ideal gases (Ideale Gase)

5.1 Equation of state for an ideal gas (Zustandsgleichung eines idealen Gases)

The thermodynamic states of a constant mass of gas can be represented by *two of the three* coordinates P,V,T. When two of these coordinates are fixed, the third one is also fixed. This is because of the existence of a relationship between the three coordinates called an equation of state (see p 94). The behaviour of gases has been studied over a long period of time, and the gas laws (due to Boyle and Marriotte) and also due to (Charles and Gay-Lussac) can be combined to give a single equation of state.

$$PV = mR_i T$$

$$Pv = R_i T \qquad \text{where } v = V/m \text{ is the specific volume}$$

Here R_i is a constant corresponding to 1 kg of a gas. Its value *depends* on the *particular gas* and its units are J / kg K.

If we consider a *mole* of a gas, the value of the constant is *the same for all gases*. The constant used in this case is designated R and is known as the *molar universal gas constant*. It has a value of $R = 8.3144$ J / mol. K and we can write

$$PV = nRT \qquad \text{where } n \text{ is the number of moles}$$

The above equation of state holds not only for ideal gases, but also for *real gases* when the *pressure approaches zero*. Experiments show that the behaviour of all real gases *approaches* that of an *ideal gas* as the pressure approaches zero.

5.2 Specific heats of gases (Spezifische Wärmekapazität der Gase)

Two types of specific heat are useful in discussing the behaviour of an ideal gas,

c_p the specific heat at constant pressure

and c_v the specific heat at constant volume

It can be shown that the difference between the specific heats is given by

$$c_p - c_v = R_i$$

The ratio of the specific heats is an important constant and is defined as

$$\frac{c_p}{c_v} = \kappa \text{ (or sometimes termed } \gamma)$$

The value of the constant κ depends on the *atomicity* of the gas, and is a function of the *number* of *degrees of freedom* of the molecules of the gas. Values corresponding to different types of atomicity are given below.

Atomicity of the gas	Degrees of freedom	c_p	c_v	c_p/c_v
1. Monatomic	3	$\frac{5}{2}R_i$	$\frac{3}{2}R_i$	$\frac{5}{3}$
2. Diatomic	5	$\frac{7}{2}R_i$	$\frac{5}{2}R_i$	$\frac{7}{5}$
3. Polyatomic	n	$\frac{n+2}{2}R_i$	$\frac{n}{2}R_i$	$1+\frac{2}{n}$

5.3 Adiabatic (isentropic) process for an ideal gas
(Isentrope Zustandsänderung eines idealen Gases)

It can be shown for an ideal gas quasi-static adiabatic process

$$\frac{P_1}{P_2} = \left(\frac{V_2}{V_1}\right)^{\kappa} = \left(\frac{T_1}{T_2}\right)^{\frac{\kappa}{\kappa-1}}$$

It can also be shown that

(a) the slope of an isothermal curve is $\quad \left(\dfrac{\partial P}{\partial V}\right)_T = -\dfrac{P}{V}$

(b) the slope of an adiabatic curve is $\quad \left(\dfrac{\partial P}{\partial V}\right)_S = -\kappa\dfrac{P}{V}$

5.4 Changes of state of an ideal gas
(Zustandsänderungen eines idealen Gases)

A change of state of a gas involves a change in its thermodynamic coordinates. It is useful to consider changes of a *specific type*, and some such changes corresponding to unit mass (1 kg) of the gas are discussed below. It is convenient to represent these changes on P,v and T,s diagrams.

5.4.1 Constant volume (isochoric) change (Isochore Zustandsänderung)

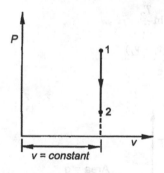
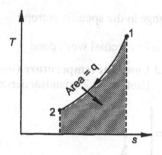

Fig 4.8. (a) P vs v diagram	(b) T vs s diagram

When the volume is constant $\qquad \dfrac{P_1}{P_2} = \dfrac{T_1}{T_2}$

Heat gained or lost $\qquad q = c_v(T_2 - T_1)$

Since there is no change in volume $\qquad \int P dv = 0$.

Change in the specific internal energy $\qquad \Delta u = c_v(T_2 - T_1)$

Change in the specific enthalpy is $\qquad \Delta h = c_p(T_2 - T_1)$

Change in the specific entropy is $\qquad \Delta s = c_v \ln\dfrac{T_2}{T_1}$

The external work done is $\qquad \int P dv = 0$.

5.4.2 Constant pressure (isobaric) change (Isobare Zustandsänderung)

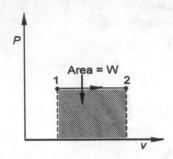

Fig 4.9 (a) P vs v diagram   (b) T vs s diagram

When the pressure is constant $\qquad\qquad \dfrac{v_1}{v_2} = \dfrac{T_1}{T_2}$

Heat gained or lost $\qquad\qquad\qquad\quad q = c_P\,(T_2 - T_1)$

Change in the specific internal energy $\quad \Delta u = c_v\,(T_2 - T_1)$

Change in the specific enthalpy $\qquad\quad \Delta h = c_P\,(T_2 - T_1)$

Change in the specific entropy $\qquad\quad \Delta s = c_P \ln\dfrac{T_2}{T_1}$

Specific external work done $\qquad\qquad W = P(v_1 - v_2)$

5.4.3 Constant temperature (isothermal) change
      (Isotherme Zustandsänderung)

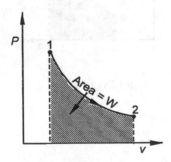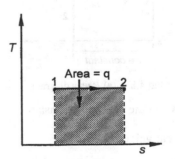

Fig 4.10   (a) P vs v diagram   (b) T vs s diagram

When the temperature is constant $\qquad\qquad \dfrac{P_1}{P_2} = \dfrac{v_2}{v_1}$

Heat gained or lost $\qquad\qquad\qquad\qquad q = R_i T \ln\dfrac{v_2}{v_1} = R_i T \ln\dfrac{P_1}{P_2}$

Change in specific internal energy $\qquad\quad \Delta u = 0$

Change in specific enthalpy $\Delta h = 0$

Change in specific entropy $\Delta s = R_i \ln \dfrac{v_2}{v_1} = R_i \ln \dfrac{P_1}{P_2}$

Specific external work done $W = R_i T \ln \dfrac{v_1}{v_2} = R_i T \ln \dfrac{P_2}{P_1}$

5.4.4 Adiabatic (isentropic) change (Isentrope Zustandsänderung)

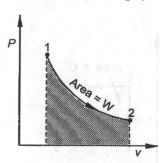

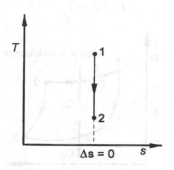

Fig 4.11 (a) P vs v diagram   (b) T vs s diagram

$$\frac{P_1}{P_2} = \left(\frac{v_2}{v_1}\right)^{\kappa} = \left(\frac{T_1}{T_2}\right)^{\frac{\kappa}{\kappa-1}}$$

When an adiabatic change takes place, there is no transfer of heat between a system and its surroundings.

The heat change is zero $q = 0$

Work done $W = \Delta u = c_v (T_2 - T_1)$

Since $q = 0$, Change in specific entropy $\Delta s = 0$

Change in the specific enthalpy $\Delta h = c_p (T_2 - T_1) = \dfrac{\kappa}{\kappa - 1} P_1 v_1 \left(\dfrac{T_2}{T_1} - 1\right)$

5.5 Cyclic processes and heat engines
(Kreisprozesse und Wärmekraftmaschinen)

Several types of heat engines are used in practice to convert *heat energy* into *work*. In the analysis of these cycles it is assumed that friction, turbulence, etc. *can be neglected*. Although this is not true in practice, the *conclusions* from such studies are *approximately valid* and are used in *evaluating the performance* of heat engines.

5.6 The Carnot cycle (Der Carnot-Prozess)

The Carnot cycle is a *reversible cycle* of *historical importance* in that a Carnot engine has the *highest efficiency* of all heat engines working between two heat reservoirs at different temperatures. In practice all *practical heat engines* have *efficiencies* which are *well below* that of an ideal Carnot engine. The stages in

the cycle are shown in the diagrams of Fig 4.12 (a) and (b). The Carnot cycle consists of *two isothermal* and *two adiabatic* (isentropic) stages.
The stages in the cycle are as follows:

$1 \rightarrow 2$ Isothermal expansion at temperature T_1 ($= T_2$)
$2 \rightarrow 3$ Adiabatic expansion resulting in a fall of temperature from T_1 to T_3
$3 \rightarrow 4$ Isothermal compression at temperature T_3 ($= T_4$)
$4 \rightarrow 1$ Adiabatic compression resulting in an increase of temperature from
         T_3 ($= T_4$) to T_1

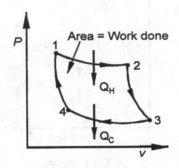
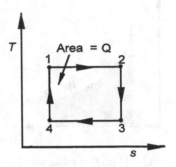

Fig 4.12 Carnot cycle (a) P vs v diagram   (b) T vs s diagram

The Carnot cycle is a *reversible cycle*. It can be shown that the heat absorbed from the hot reservoir Q_{12} ($= Q_H$) at a temperature T_1 and the heat rejected into the cold reservoir Q_{34} ($= Q_C$) at a temperature T_2, are in the *ratio of their absolute temperatures*. This means that

$$\frac{Q_{12}}{Q_{34}} = \frac{T_1}{T_3}$$

$$W = Q_{12} - Q_{34}$$

Work done $$W = Q_{12}\left(1 - \frac{Q_{34}}{Q_{12}}\right)$$

$$W = Q_{12}\left(1 - \frac{T_3}{T_1}\right)$$

Efficiency $$\eta = \frac{W}{Q_{12}}$$

$$\eta = \left(1 - \frac{T_3}{T_1}\right)$$

The efficiency of a Carnot engine depends *only on the temperatures* of the reservoirs. This is an *idealized cycle* which cannot be realized in practice. However its usefulness lies in the fact that it gives a value for the **upper limit** of *efficiency* that a heat engine can reach.

5.7 The Otto cycle (Der Otto-Prozess)

The Otto cycle is mainly used in internal combustion *petrol engines*. In these engines, a mixture of petrol and air are *compressed* and *ignited,* to produce the *power* which is required to drive the engine.

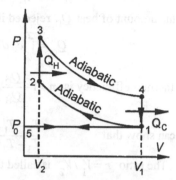

An analysis of the cycle can only be made by making simplifying assumptions. The working substance is assumed to be like an ideal gas, and all processes assumed to be quasi-static. Friction is assumed to be negligible.

Fig 4.13 The Otto cycle

The Otto cycle consists of six processes. *Four* of these involve a *motion of the piston* and are called *strokes*. The processes are described briefly below.

1. The **induction** (or intake) **stroke** $5 \rightarrow 1$, in which a mixture of gasoline vapour and air is sucked into the cylinder at atmospheric pressure P_0 due to the *downward movement* of the piston.

2. The **compression stroke** $1 \rightarrow 2$, in which the piston *moves upwards* to compress the mixture (adiabatically). This causes a considerable rise in temperature. The temperature rises from T_1 to T_2.

3. The **ignition process** $2 \rightarrow 3$ represents a constant volume (isochoric) increase of temperature. This is brought about by the *absorption* of a quantity of *heat* Q_H which is produced by the explosive *combustion* of the *fuel*. The piston *remains stationary* during this process while the temperature rises from T_2 to T_3.

4. The **power stroke** $3 \rightarrow 4$ is a consequence of the hot gases expanding and pushing the *piston downwards*. This is an adiabatic expansion which involves a drop in temperature from T_3 to T_4.

5. The **valve exhaust process** $4 \rightarrow 1$ in which the *exhaust valve opens* and allows some *gas to escape*. The pressure drops to the atmospheric pressure value of P_0. The piston does not move during this process. This represents a constant volume (isochoric) drop in temperature from T_4 to T_1 and a rejection of an amount of heat Q_C.

6. The **exhaust stroke** $1 \rightarrow 5$ involves the *movement of the piston* pushing out all the remaining gases out of the cylinder. This represents a constant pressure (isobaric) process at atmospheric pressure. The volume changes from V_1 to zero.

The amount of heat Q_H absorbed in the ignition process $2 \rightarrow 3$ is given by

$$Q_H = \int_{T_2}^{T_3} c_V \, dT = c_V (T_3 - T_2)$$

and the amount of heat Q_C rejected in the exhaust process $4 \rightarrow 1$ is given by

$$Q_C = \int_{T_4}^{T_1} c_V \, dT = c_V (T_4 - T_1)$$

The thermal efficiency is $\eta = 1 - \dfrac{Q_C}{Q_H} = 1 - \left(\dfrac{T_4 - T_1}{T_3 - T_2} \right)$

We can show that $\dfrac{T_4 - T_1}{T_3 - T_2} = \left(\dfrac{V_2}{V_1} \right)^{\kappa - 1}$

The ratio $r = V_1 / V_2$ is called the compression ratio and we can write

$$\eta = 1 - \frac{1}{(V_1 / V_2)^{\kappa - 1}} = 1 - \frac{1}{r^{\kappa - 1}}$$

5.8 The Diesel cycle
(Der Diesel-Prozess)

In the Diesel cycle, *only air* is initially *sucked in*. This air is *compressed* adiabatically until the temperature of the air is *high enough* to *ignite* the fuel, which is *sprayed* directly into the combustion chamber of the cylinder. The successive stages in the Diesel cycle are described below. The rate of supply of the fuel can be adjusted to control the rate of combustion.

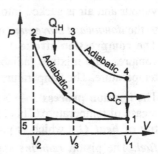

Fig 4.14 The Diesel cycle

1. The *induction* (or intake) *stroke* $5 \rightarrow 1$. Only air is sucked in.
2. The *compression stroke* $1 \rightarrow 2$. This involves the *adiabatic compression* of air to a temperature that is high enough to *ignite* the fuel *sprayed* into the cylinder after the compression.
3. The *combustion stroke* $2 \rightarrow 3$. The fuel is *sprayed in* at *such a rate* that the piston *moves out* during the combustion process which takes place at *constant pressure* (isobaric).

The remainder of the cycle, power stroke, valve exhaust and exhaust stroke is the same as for the gasoline engine.

If we write, expansion ratio $r_E = \dfrac{V_1}{V_3}$ and compression ratio $r_C = \dfrac{V_1}{V_2}$

then it can be shown that $\eta = 1 - \dfrac{1}{\kappa} \dfrac{(1/r_E)^{\kappa} - (1/r_C)^{\kappa}}{(1/r_E) - (1/r_C)}$

6 The transfer of heat (Wärmeübertragung)

Heat is *energy in transit*, and can flow from one part of a system to another, or from a system to its surroundings. Heat transfer can take place in *three different ways*, (1) conduction (2) convection (3) radiation.

6.1 Heat conduction (Wärmeleitung)

Consider a thin slab of material which has a thickness Δx and a surface area A. Let one surface of the slab be maintained at a temperature θ while the other surface is kept at a temperature $\theta + \Delta\theta$. If the quantity of heat Q that flows perpendicular to the surfaces for a time τ is measured, then it is found that

The rate of flow of heat $\dfrac{Q}{\tau} \propto A \dfrac{\Delta\theta}{\Delta x}$

For a slab of infinitesimal thickness, we can write

$$\frac{dQ}{dt} = -\kappa A \frac{d\theta}{dx}$$

The derivative $d\theta / dx$ is called the *temperature gradient*. The negative sign is used to ensure that the direction of heat flow should be coincident with the positive direction of the x axis. This type of heat transfer is called *heat conduction* and the constant κ which is called the *thermal conductivity* has a value which depends on the material.

A substance which has a large value of κ is called a *thermal conductor*, while one with a small value of κ is called a *thermal insulator*. The values of κ can be measured for different substances for different ranges of temperature.

6.2 Heat convection (Wärmeübergang/ Wärmekonvektion)

The transfer of heat by convection takes place in *liquids* or *gases*. When a fluid is heated, some parts of the fluid acquire a different density in comparison with other parts of the fluid, and a current (of fluid) flows due to the difference in density. Such a current is called a *convection current*. A convection current absorbs heat in one part of a fluid and moves it to a cooler part of the fluid where it rejects the heat.

6.3 Thermal radiation (Wärmestrahlung)

Thermal radiation is the term used to designate the *radiation emitted* by a body by *virtue of its temperature*. The radiation consists of a *continuous spectrum* of electromagnetic waves. The *total energy* radiated annd the *distribution of energy* with wavelength are dependent on the temperature. As the temperature of a body is increased, the total energy radiated also increases, and the wavelength at which the *maximum* amount of energy is emitted *becomes shorter*.

The rate at which thermal radiation is emitted by a body depends on the *temperature* and the nature of its *surface*. The total radiant power emitted per unit area of a surface is called the *emissive power* of a surface. (Other names like emittance, radiant flux density, or radiant excitance have also been given for the emissive power). The units are kW / m^2 at a given temperature.

When thermal radiation falls on a body equally from all directions (which means isotropically), some of it is *absorbed*, some *reflected* and some *transmitted*. The *fraction absorbed* is called the *absorptivity* α (or the absorptance). The magnitude of this fraction depends on the temperature and the nature of the surface.

Emissive power	R = Total radiant power emitted per unit area
Absorptivity	α = Fraction of isotropic radiation which is absorbed

6.4 Black body radiation and cavity radiation
(Schwarzkörperstrahlung und Hohlraumstrahlung)

An ideal body which *absorbs all thermal radiation* falling on it $(\alpha = 1)$ is called a black body. In practice some substances like *lamp black* (Lampenruß)have an absorptivity of nearly unity.

A *hollow cavity* (or box) with a small hole in one of its walls is a *very good approximation* to a black body. Any radiation entering the cavity through the hole is partly absorbed and partly reflected many times by the interior walls of the cavity. Only a very *minute fraction* of this radiation can escape through the hole again. This takes place *regardless* of the type of *material* used for the walls. The result is that the cavity absorbs all the radiation falling on it, and the radiation inside it is *isotropic*. If a small amount of radiation is allowed to escape through the small hole in the cavity, this radiation is independent of the material used for the interior walls and depends *only on the temperature* of the cavity. This radiation called *cavity radiation*, can be assumed to be the *same* as that from a *black body* maintained at the same temperature as the cavity.

6.5 Kirchhoff's law (Kirchhoffsches Gesetz)

If the *emissive power of a black body* is R_B, then the emissive power of a *nonblack* body R is a *fraction* of this. According to Kirchoff's law, this fraction is equal to the absorptivity of the body. Therefore we can write

$$R = \alpha R_B$$

6.6 The Stefan-Boltzmann law (Gesetz von Stefan und Boltzmann)

On the basis of experimental evidence, it was stated by Stefan in 1879 that the heat transferred by radiation between a body and its surroundings was proportional to the fourth power of the absolute temperature. This was derived theoretically by Boltzmann, who was able to show that the emissive power of a black body at any temperature T is equal to

$$R_B = \sigma T^4$$

This is known as the Stefan-Boltzmann law and the constant σ is called the Stefan-Boltzmann constant. The constant has a value

$$\sigma = 5.67 \times 10^{-8} \ W/m^2(K)^4$$

V Machine elements (Maschinenelemente)

1 Limits and fits (Grenzmaße und Passungen)

1.1 Measurement and inspection (Messen und Lehren)

In the past, it was the practice to make each component to *precise dimensions* and assemble the components to form the final product. Each *dimension* of the component was *measured* accurately using measuring instruments, and accepted only when the dimensions were extremely *close* to the *prescribed dimensions*.

With the advent of *mass production*, it was no longer possible to manufacture and measure each component to obtain an *exact fit*. The dimensions of mass produced components varied from sample to sample and the *measurement* procedure had to be *replaced* by a different procedure called *inspection*. The main feature of this inspection procedure was to check if each dimension of a sample lay between *two prescribed limits*, an *upper limit* and a *lower limit*.

The inspection procedure is *simpler* than the measurement procedure and can be accomplished for example by using *limit gauges*. These gauges ensure that the dimensions of a component always lie between *prescribed limits*.

This procedure guaranteed the *interchangeability* of components, regardless of when and where the components were produced. Soon *international standards* became *desirable* and *necessary,* and it has become the practice for manufacturers to adopt the **ISO Standards** detailed below in Section 1.3.

1.2 Basic quantities (Grundbegriffe)

Some of the basic quantities used when choosing suitable limits and fits for *cylindrical holes* and *shafts* are defined below and illustrated in Fig 5. 1.

1. **Basic size** (Nennmaß) is the *theoretical size* from which limits or deviations are measured. The basic size is the *same for both members* (shaft and hole).
2. **Upper (or high) limit** (Höchstmaß) refers to the *maximum size* that a dimension in a component can have.
3. **Lower (or low) limit** (Mindestmaß) refers to the *minimum size* that a dimension in a component can have.
4. **Deviation** (Abmaß) refers to the algebraic difference between a *given size* and the *corresponding basic size*.
5. **Upper deviation** (oberes Abmaß) is the algebraic difference between the *upper (or high) limit* and the *corresponding basic size*.
6. **Lower deviation** (unteres Abmaß) is the algebraic difference between the *lower (or low) limit* and the *corresponding basic size*.
7. **Tolerance** (Toleranz) is the difference between the *upper* and *lower limits* of the dimensions of a component.
8. **Fundamental deviation** (Grundabmaß) is distance of the *tolerance zone* from the *basic size*. This will be seen to be the distance of the lower or the upper limits from the basic size, *whichever lies closer*.

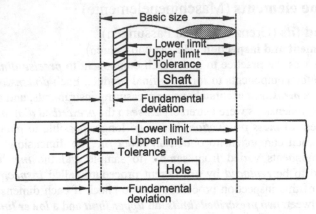

Fig 5.1. Basic quantities related to the fit of a shaft and a hole

1.3 ISO standards for limits and fits (ISO Toleranzsysteme)

The ISO standards are based on the two following items:

(1) Fundamental tolerance (2) Fundamental deviation

1.3.1 Fundamental tolerance grades (Grundtoleranzgrade)

The fundamental tolerance is specified in terms of 20 *tolerance grades*. Each tolerance grade has a *number* assigned to it. The numbers assigned are 01, 0, and 1 to 18. The actual *magnitude of the tolerance* depends on both the *tolerance grade* and the *basic size*. The basic sizes vary from 1mm to 3150 mm and are divided into 21 groups.

The tolerance grade which is to be used can be *chosen freely* by the designer depending on the *accuracy* to which the work has to be carried out. The smaller numbers correspond to *smaller tolerances*, while the larger numbers correspond to *larger tolerances* assuming that the basic size remains the same. Tolerances on components should be chosen to be *as large as possible*. This is because small tolerances require expensive manufacturing and measuring equipment, and lead to a higher percentage of *rejected components*. The tolerance grades which are suitable for different types of applications are shown in Fig 5.2.

Tolerance grades	01 to 4	5 to 11	12 to 18
Type of application	Test gauges, Standards, Instruments	Machine tools, Manufacture of vehicles	Ordinary machines, Consumer goods
Processes used	Lapping, Honing, Superfinishing	Turning, Milling, Grinding	Pressing, Drawing, Forging

Fig 5.2. Tolerance grades suitable for different applications

1.3.2 Fundamental deviation (Grundabmaß)

The fundamental deviation determines the *type of fit* obtained when a shaft is mated to a hole. If we have a hole that is close to the basic size, then the *greater* the fundamental deviation of the shaft, the *coarser* will be the fit between hole and shaft. The fundamental deviations are indicated by the following letters:

For holes: A B C D E F G H J JS K M N P R S T U V W X Y Z ZA ZB ZC
For shafts: a b c d e f g h j js k m n p r s t u v w x y z za zb zc

The fundamental deviation is *different* for each of these letters and is illustrated in Fig 5.3. The letters JS for holes and js for shafts correspond to tolerance boundaries which are symmetrical relative to the zero line.

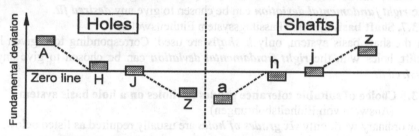

Fig 5.3 Position of the tolerance boundaries for holes and shafts

1. A *hole* is described by an appropriate capital letter followed by a number denoting the tolerance grade   e.g H7
2. A *shaft* is described by an appropriate small letter followed by a number denoting a tolerance grade   e.g. p6
3. A *fit* is described by writing the hole symbol followed by the symbol for the shaft   e.g. H7/p6

1.3.3 Types of fits (Passungsarten)

1. **Fit (Passung)** – The term fit refers to the *difference* between the *size of the hole* and the *size of the shaft* when both members have the same basic size.
2. **Clearance fit (Spielpassung)** – A clearance fit is obtained when the *low limit* of a *hole* exceeds the *high limit* of a *shaft* which is to mate with the hole.
3. **Interference fit (Übermaßpassung)** – An interference fit is obtained when the high limit of the hole is *smaller* than the low limit of the shaft.
4. **Transition fit (Übergangspassung)** – In a transition fit there can be either a *clearance* or an *interference* between shaft and hole. In practice the tolerances for transition fits are *very small,* and both the hole and shaft are around the middle limit. Any interference that exists will be slight, and *hand pressure* is usually sufficient to push the shaft into the hole.

1.3.4 Systems of fits (Passungssysteme)

In order to keep *manufacturing* and *inspection costs low*, industry has largely adopted either a *hole basis system* (based on a constant hole size) or a *shaft basis system* (based on a constant shaft size).

1.3.5 Fundamental deviations for H holes and h shafts

(Grundabmaße für H Bohrungen und h Wellen)

All H holes and h shafts have zero deviation. The *lower limit* for H holes is the same as the *basic size*. For shafts, the *upper limit* is equal to the *basic size*.

1.3.6 Hole basis system (Passungssystem Einheitsbohrung)

In the hole basis system only *H holes* are used. As mentioned above, the *fundamental deviation* for all H holes is *zero*. For a given H hole, *shafts* with the *right fundamental deviation* can be chosen to give any *desired fit*.

1.3.7. Shaft basis system (Passungssystem Einheitswelle)

In the shaft basis system, only *h shafts* are used. Corresponding to a given h shaft, holes with the *right fundamental deviation* can be chosen to give any *desired fit*.

1.3.8 Choice of suitable tolerance grades for holes on a hole basis system

(Auswahl von Einheitsbohrungen)

For ordinary work only *six grades of holes* are usually required as listed below:

H6	Internal grinding or honing	H9	Boring with a worn lathe
H7	High quality boring, broaching	H10	Good quality drilling
H8	Boring with a lathe, reaming	H11	Standard drilling

1.3.9 Some preferred fits using the hole basis system (Auswahl von
    (a) Clearance fits   Paßtoleranzfeldern)

1. Loose running fit	H7/d8, H8/d10, H11/d11
2. Easy running fit	H6/e7, H7/e8, H8/e9
3. Running fit	H6/f6, H7/f7, H8/f8
4. Close running fit	H6/g5, H7/g6, H8/g7
5. Location fit (not for running)	H6/h5, H7/h6, H8/h7

(b) Transition fit		**(c) Interference fit**	
1. Push fit	H6/j5, H7/j6	1. Light press fit	H6/p5. H7/p6
2. Easy keying fit	H6/k5, H7/k6	2. Press fit	H6/s5, H7/s6
3. Drive fit	H7/n6, H8/n7	3. Shrink fit	H6/u5, H7/u6

2 Rivets and riveted joints (Niete und Nietverbindungen)

Riveting is mainly used when it is necessary to join two or more metal sheets (or other components) **permanently**. Although it has been **replaced** to a **large extent** by **welding**, it has however many **advantages** over welding. Among these are that the

- microstructure of the metal remains unchanged
- it is possible to join different types of materials.
- on-site riveting is possible.
- process is easily controlled.

  Disadvantages are that:
- the material is weakened by holes, weak joints,etc.
- working times are longer

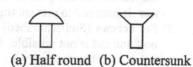
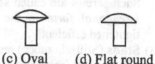

(a) Half round (b) Countersunk

(c) Oval   (d) Flat round

Fig 5.4 Some types of rivet heads

Riveting is still used to make firm and **leakproof joints** in ships, aircraft, steel containers, boilers, etc. In addition to its use in joining steel sheets, it is also used to join materials like copper, aluminium and their alloys.

2.1 Types of rivets (Nietformen)

Some of the types of rivets used are shown in Fig 5.4. Rivets with a **half round** head are the **most frequently** used, but **countersunk rivets** must be used for joints that need to have a **flush surface**.

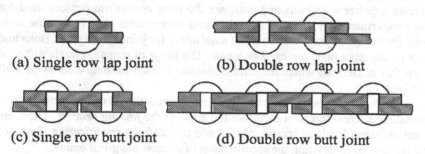

(a) Single row lap joint   (b) Double row lap joint

(c) Single row butt joint   (d) Double row butt joint

Fig 5.5 Some types of riveted joints

2.2 Types of riveted joints (Nietverbindunsarten)

Two types of joints are usually used, **lap joints** and **butt joints**. Lap joints are used in boiler and container construction, while butt joints are used in steel construction. Some examples of the types of joints used are shown in Fig 5.5.

3 Screws and screw joints (Schrauben und Schraubenverbindungen)
Metal sheets and components can be joined by screws in several different ways
as shown in Fig 5.6.

1) **Bolts and nuts** (Schrauben und Muttern) are used when both sides of
   the components to be joined are *accessible* (Fig 5.6(a)). If the parts are
   subject to vibration, an additional part like a *spring washer* or a *lock
   nut* is required to prevent the nut from becoming loose.

2) **Set screws** (Stellschrauben) are used (Fig 5.6(b)) when the use of a
   bolt and nut is not possible. Set screws with normal heads can be used,
   but it is often necessary to use set screws with *countersunk heads*.
   Such screws are called *socket screws* (Fig 5.6(c)) and have a hole of
   hexagonal form in the head of the screw, enabling them to be
   tightened efficiently.

3) **Studs** (Stiftschrauben) are used for example (Fig 5.6(d)), when joining
   parts to *cast iron components*. Cast iron has a low tensile strength and
   excessive tightening of a set screw into a cast iron thread can cause the
   thread to be damaged. In this case the studs are screwed into the
   casting first, and the tightening is accomplished by using *mild steel
   nuts*. Any *damage* caused will be to the nut or stud and not to the
   casting. Studs can be used to effect *gas-tight* and *water-tight* joints in
   cases when *heavy pressures* are present. A good example of the use of
   studs is their use for holding down a *cylinder head* on a *cylinder block*
   of an internal combustion engine. A *thin gasket* is placed between the
   metal surfaces to make the joint gas and water-tight.

3.1 Screws, bolts and nuts (Schrauben und Muttern)
Screws together with bolts and nuts, are the most convenient devices used for
the *nonpermanent* joining of materials and components. A screw is the term
used for a device (like a wood screw) *used alone* for joining two parts. Bolts and
nuts on the other hand are *used as a pair*. The usage of terms is clearly different
from that in German where the term *Schraube* is used for both a screw as well
as a bolt.
A screw joint is made by screwing an external screw thread on an internal screw
thread. The screw thread is therefore the *basis* of the *joining process*. There are
many different kinds of screws, each made to a *definite specification*. Some of
the terms used in defining the specification of a screw are given below.

1) The angle of a screw thread is the angle between the *two inclined faces* of
   the screw thread Fig 5.7 (a).

2) The pitch is the distance measured between *any point* on a thread and the
   corresponding point on the next thread when measured parallel to the
   axis of the screw Fig 5.7 (a).

3) The major diameter is the *external diameter* of the screw, and the minor
   diameter is the core or *smallest diameter* of the screw.

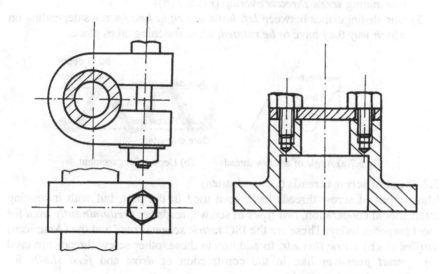

Fig 5.6 (a) Use of bolts and nuts   Fig 5.6(b) Use of set screws

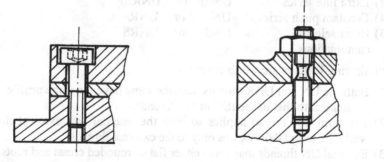
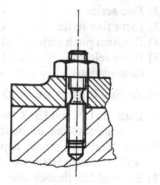

Fig 5.6 (c) Use of socket screws   Fig 5.6 (d) Use of studs

4) The depth of engagement is the radially measured distance over which the two mating *screw threads overlap* (Fig 5.7 (b)).

5) One distinguishes between *left hand* and *right hand* screws depending on *which way* they have *to be rotated*, when fastening takes place.

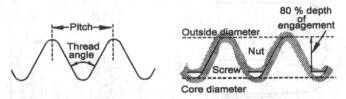

Fig 5.7(a) Angle of a screw thread   (b) Depth of engagement

3.2 Types of screw threads (Gewindearten)

Many types of screw threads have been used in the past, but with increasing international cooperation, *two types* of screw threads are *predominantly used* for most purposes today. These are the ISO *metric screw thread* and the (American) *unified* (inch) *screw threads*. In addition to these, other screw threads are used for *special purposes* like in the construction of *drive* and *feed shafts* for machines and machine tools.

3.2.1 Unified screw threads ((American) Unified Gewinde)

These screw threads are mainly used in the U.S and Canada. The *basic profile* of the unified screw thread is the same as that for the ISO metric thread. The series of unified threads that are available are:

1) Coarse series   UNC   or   UNRC
2) Fine series   UNF   or   UNRF
3) Extra fine series   UNEF   or   UNREF
4) Constant pitch series   UN   or   UNR
5) Other selected   UNS   or   UNRS
   combinations

The following features are worth noting.

1) Both the UN and UNR threads have the same have the same profile which is identical to that of the ISO metric threads.
2) The term UN thread applies to both the internal and external threads, while the term UNR applies only to the external threads.
3) External UN threads may have either flat or rounded crests and roots.
4) Internal UN threads must have rounded roots, but may have flat or rounded crests.
5) Internal UN threads must have rounded roots.

3.2.2. ISO metric threads (Metrisches ISO Gewinde)

As mentioned before, the profile of the ISO metric thread is the same as the unified thread. There are a number of metric thread series, some of which are mentioned below.

1) **ISO metric series** – Covers thread diameters from 1mm to 68 mm. Intended for use in all types of bolts and nuts and other types of mechanical fasteners.

2) **ISO metric fine thread series** – Covers thread diameters from 1mm to 300mm. Used for mechanical fasteners, for ensuring *water-tight* and *gas-tight* joints, for measuring instruments, etc.

3) **ISO metric saw tooth thread** – The saw tooth thread has a thread angle of 33°. The unsymmetrical thread form makes *unsymmetrical loading* possible. Used in the construction of collett chucks for lathes and milling machines.

4) **ISO metric acme thread** (trapezoidal form) – Covers thread diameters from 8mm to 300mm. Used in drives for machine tools, vices, valves, presses, etc.

3.2.3 Other types of threads (Andere Gewindearten)

1) **Whitworth pipe threads** – This thread has an angle of 55 degrees and is used in pipes and pipe parts where *effective sealing* is important.

2) **Round screw thread** – These threads have a thread angle of 35 degrees together with *rounded roots* and *crests*. Used for example in clutch spindles.

3.2.4 Types of screws, bolts and nuts (Schraubenarten und Mutterarten)

1) **Different types of heads** - In addition to having different types of screw threads, screws can also have different types of heads. Some of the types of heads which are used are shown in Fig 5.8.

2) **Taper screws** – Screws which have a *tapering form* are not used with a nut. Examples of these are *wood screws* and hardened *self-tapping screws* used to join metals. The self-tapping screws are able to cut a thread in the metal when they are screwed in.

3) **Nuts** – Nuts can also have different forms. Some of the available forms are shown in Fig 5.9.

4) **Thread inserts** – Thread inserts are used with materials like soft metals, plastics and wood where the thread *strips off* or is *easily damaged*. They can also be used to repair damaged screw threads.

5) **Locking devices** – Locking devices are often necessary to prevent screws and bolts from becoming loose. Some of the devices that can be used are own in Fig 5.10.

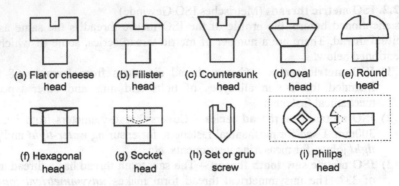

(a) Flat or cheese head   (b) Filister head   (c) Countersunk head   (d) Oval head   (e) Round head

(f) Hexagonal head   (g) Socket head   (h) Set or grub screw   (i) Philips head

Fig 5.8 Some types of screw heads

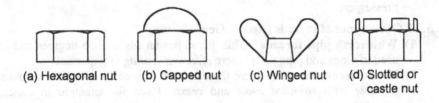

(a) Hexagonal nut   (b) Capped nut   (c) Winged nut   (d) Slotted or castle nut

Fig 5.9 Some types of nuts

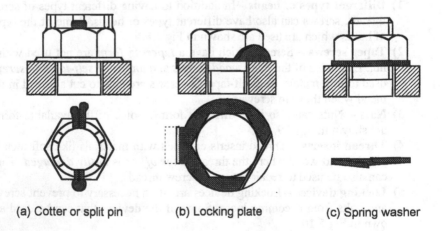

(a) Cotter or split pin   (b) Locking plate   (c) Spring washer

Fig 5.10 Some types of locking devices

4 Pins (Stifte)

4.1 Uses of pins (Verwendung von Stiften)

Pins are *removable fasteners*. They are used are used as *locating devices* and as fasteners for the *transmission* of *small torques*.

4.1.1 Locating pins (Paßstifte)

Locating pins are used to locate (or secure) the *position* of two components relative to each other. They facilitate the *precise assembly* of components and prevent *sideways movement* due to lateral (sideways) forces.

4.1.2 Fastening pins (Befestigungsstifte)

Fastening pins are used to hold components together firmly so that they can *transmit forces* and *couples* without becoming loose.

4.1.3 Overload protection pins (Abscherstifte)

Overload protection pins are used to *prevent damage* when components are subjected to *excessive forces* or *torques*. If the forces or torques exceed certain values, the *pin breaks* thereby ensuring that no damage is caused.

4.2 Types of pins (Stiftformen)

Pins can be classified according to their *shape* or *form* as cylindrical pins, taper pins, roll pins, spiral pins, grooved pins, etc.

4.2.1 Cylindrical pins (Zylinderstifte)

Cylindrical pins are used as *locating pins* to join parts when *strength* and *accuracy* are important, and when the parts that are joined need to be rarely separated.

4.2.2 Taper pins (Kegelstifte)

Taper pins are usually made with a standard taper of 1: 50. The hole is usually *bored* to the smallest diameter of the pin and then *enlarged* with a *taper reamer* until the pin projects 4 mm above the hole when inserted by hand. The pin is then hammered in until it is firmly fixed in the hole. The pin is *elastically deformed* in this process and creates a strong joint which however is not strong enough to resist shocks.

4.2.3 Roll pins (Spannstifte)

These are *hollow cylindrical pins* which have a *slit* along their length. They are made of spring steel and heat treated before use. The outer diameter of the pin is larger than the hole and becomes *compressed* when driven into the hole. These can be used to join components and to ensure resistance against lateral forces.

4.2.4 Spiral pins (Spiral-Spannstifte)

These are rolled in the form of a *spiral cylinder* from heat treated spring steel The outer diameter is slightly larger than the hole. The pins are rolled elastically tighter when driven into the hole. These pins (due to their elastic properties) are particularly suitable for use in joints which are *subjected to shocks*.

4.2.5 Grooved pins (Kerbstifte)

Grooved pins usually have *three grooves* along part or the whole of their length. These are used in joints where *great accuracy* is *not required*, and where the joints are rarely separated.

5 Axles and shafts (Axen und Wellen)

5.1 Axles (Axen)

Axles are used as *mountings* and *supports* for wheels, pulleys, levers etc. and are mainly subjected to *bending loads*. They are *not used* for the *transmission* of *torques*. Axles can be used in a *fixed position* as for example in cranes. They can also be used as *moving components* as for example in trains and other vehicles.

5.2 Shafts (Wellen)

Shafts are rotating machine elements which carry gear wheels, pulleys, couplings, etc. They are used to transmit torques and are subjected to both *bending* and *torsional* stresses. Shafts are of different types like for example fixed rigid shafts, shafts with joints in them and flexible shafts.

5.2.1 Rigid shafts (Starre Wellen)

These can be of many types like for example straight shafts, shafts with *offsets* in them like crankshafts, uniform shafts or shafts with *reduced cross-section* in certain places on the shaft. Shafts in machine tools called *spindles* are *often hollow* to accommodate chucks, tools, workpieces, etc.

5.2.2 Crankshafts (Kurbelwellen)

These are used to convert *reciprocating motion* into *rotary motion* as for example in *engines* or *compressors* with pistons in them.

5.2.3 Drive shafts (Getriebewellen)

These shafts often have their cross-sections reduced in certain places to enable machine elements like gear wheels, pulleys, bearings, couplings, etc. to be *easily* and *accurately fitted* on them.

5.2.4 Jointed shafts (Gelenkwellen)

These are used when the *position* of the end of a shaft *can change* as in the drive shafts of cars. The use of *universal joints* allows complete flexibility in these cases.

5.2.5 Flexible shafts (Biegsame Wellen)

These are used with small electrical devices which are fitted with high speed low torque motors. They are particularly useful when the position of the device driven by the motor (like a drill or a speedometer) *changes its position* often relative to the motor. The shafts are usually made of *several strands* of *steel wire* interwoven to give maximum flexibility. The interwoven strands are then protected by covering them with a *metal* (or other type of) *sheath*.

5.3 Shaft to hub (or collar) connections (Welle-Nabe Verbindungen)

A shaft is mainly used to *transmit rotary motion*. This is done *through machine elements* like gear wheels, pulleys, clutch plates, etc. which are mounted on the shaft. The *connection* between the *shaft* and the *machine element* which is responsible for the further transmission of the torque is called the *shaft to hub (or collar)* connection. The hub is the usually the *inner surface* (or other part) of the gear wheel or other machine element which fits on the shaft. Connections can be of two types (1) those that depend on *frictional forces* and (2) those that depend on *mechanical fastening devices*.

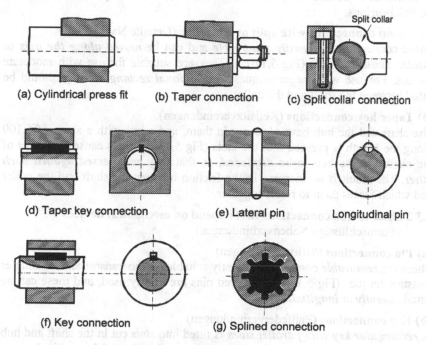

(a) Cylindrical press fit   (b) Taper connection   Split collar

(c) Split collar connection

(d) Taper key connection   (e) Lateral pin   Longitudinal pin

(f) Key connection   (g) Splined connection

Fig 5.11 Shaft to hub connections

5.3.1 Shaft to hub connections which depend on frictional forces
(Reibschlüssige Nabenverbindungen)

(a) Cylindrical press fit connections (Zylindrischer Preßverbände)

This is a *simple* and *cheap way* of making shaft to hub connections (Fig 5.11(a)). The shaft has a very slightly larger diameter than the hole into which it fits. It can be pressed into the hole if sufficient force is used. Such a fit is called an *interference fit* (see p115). The fitting can also be done by *heating the hub*, so that the hole expands allowing the hub (or collar) to be slipped easily over the shaft. On cooling the hub contracts, and grips the shaft firmly. Such connections are *permanent* and are able to transmit *large*, *variable* and *abruptly changing* torques and forces. This type of connection can be used for gear wheels, pulleys, flywheels, couplings, etc.

(b) Taper connections (Kegliger Preßverband)

These are easily removable connections in which an *outer taper* on the shaft fits into an *inner taper* in the hub (Fig 5.11(b)). They are pressed together using a nut or a screw. The *axial forces* which are brought into play give rise to *large normal forces* which hold the components together. Taper connections are capable of transmitting large, variable and abruptly changing torques. They can be used for the same applications which were mentioned for the cyindrical press connections. In addition they are also used in machine tool spindles and mounts for ball bearings.

(c) Clamp connections with split or slit collar (Geteilte Nabe)

These connections are *easily removable* and can be *moved along the axis* or *rotated about the axis* (Fig 5.11(c)). They are suitable for use with moderate torques. For use with larger torques an *additional rectangular key* should be fitted between the shaft and the collar.

(d) Taper key connections (Keilsitzverbindungen)

The shaft and the hub have *slots cut* in them, and a key with a slope of 1:100 along the length is pressed into the slots (Fig 5.11(d)). This causes the *axes* of the shaft and the hub to be *displaced*, so that they are *pressed against each other*. The result is an increase in the friction between the shaft and the collar and which forces them to rotate together.

5.3.2 Shaft to hub connections that depend on mechanical devices
         (Formschlüssige Nabenverbindungen)

(a) Pin connections (Stiftverbindungen)

These are *removable connections* mostly suitable for the transmission of smaller constant torques (Fig 5.11(e)). Tapered pins are mostly used, and these can be fitted *laterally* or *longitudinally*.

(b) Key connections (Paßfederverbindungen)

A *rectangular key* with *parallel sides* is fitted into *slots* cut in the shaft and hub (Fig 5.11(f)). There is a *space* between the key and the top of the slot in the hub. Such a connection is unsuitable for torques which are subject to *abrupt changes*. For gear wheels which have to *slide along the shaft*, the keys are made with *sufficient tolerances* to enable sliding. By *using screws* to *fix the key* on the shaft or the collar, any axial movement of the key can be avoided. A key that is fixed in this fashion is called a *feather key*.

(c) Splined connections (Profilwellen Verbindungen)

Splined connections are used for *heavy duty* couplings (Fig 5.11(g)). An axial movement between the shaft and the hub is possible when splines are used. The spline shaft has a number of *longitudinal projections* round its circumference. These engage with corresponding *recesses* in the hub. These splined shafts can have *different profiles* like involute (similar to those in gear wheels), parallel-sided slots and splines, polygon profiles,etc.

5.3.3 Axial locking devices (Wellensicherungen)
Many types of locking devices are available for preventing the *axial movements* of shafts and machine elements like ball bearings, bushes, pulleys, etc.

a) Locking rings (Sicherungsringe)
These are made of spring steel and are *round in shape*. They exert uniform pressure round the circumference of a slot or recess.

(b) Snap rings (Springringe)
These are used where a ring of *uniform cross-section* is required.

(c) Locking discs (Sicherungsscheibe)
These are used for small shafts in instruments and other devices.

(d) Cotter pins (Splint)
These are particularly useful in preventing bolts and nuts from *becoming loose*.

(e) Adjusting ring or set collar (Stellringe)
These are used to *limit the axial movement* of shafts or to *allow the sideways motion* of moving elements like wheels and levers.

6 Couplings (Kupplungen)
The main purpose of a coupling is to transmit the torque from one shaft to another shaft (or to another drive element). There are many types of couplings.

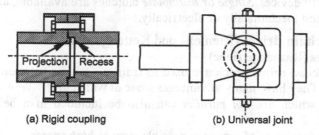

(a) Rigid coupling   (b) Universal joint

Fig 5.12 Some Couplings

6.1 Rigid couplings (Starre Kupplungen)
Rigid couplings make a *firm connection* between a shaft and another drive element. One form of rigid coupling is shown in Fig 5.12 (a). One of the shafts has a *cylindrical projection* which fits into a corresponding *cylindrical recess* on the other shaft. The shafts can be screwed together to be *friction tight*. When the coupling has to be disconnected, the shafts must be moved apart in the axial direction.

6.2 Flexible inelastic couplings (Flexible unelastische Kupplungen)
These are required to connect shafts which are *misaligned laterally* or *angularly*. Universal joints (Gelenkkupplungen) can be used where very large misalignments are present, and are particularly useful for connecting the *engine* of a vehicle to the *final drive axle*. Two such joints are usually coupled through sliding *splined drive* shafts, to allow both *longitudinal* and *lateral movement* of the axle relative to the engine. One type of universal joint is shown in Fig 5.12(b).

6.3 Elastic couplings (Elastische Kupplungen)
Elastic couplings can tolerate *considerable misalignment* in the radial and axial directions. They are also helpful in *damping* out vibrations. They are most often used in compressors and pumps where sudden variations in torque are experienced. Rubber parts and leaf springs are used as elastic components.

6.4 Clutches (Schaltbare Kupplungen)
Clutches are used when the coupling between the two shafts needs to be *engaged* and *disengaged*, even when the *shafts are rotating*.

6.4.1 Claw-type clutches (Klauenkupplungen)
These clutches have two sets of claws which *grip each other* when the clutch is engaged. No additional force is required to keep the clutch in the engaged position when the shafts are coupled together. They have square jaws when driven in both directions, or spiral jaws when driven in one direction.

6.4.2 Friction clutches (Kraftschlüssige Schaltkupplungen)
Friction clutches reduce the *coupling shock* experienced, when engagement takes place. They also slip when the torque *exceeds* a *certain value*, thereby acting as safety devices. *Single* or *multiplate* clutches are available, and they can be operated mechanically or electrically.

7 Belt and chain drives (Riemen- und Kettengetriebe)
7.1 Belt drives (Riemengetriebe)
These are *friction drives* which are used to transmit torque between two or more shafts. They have many advantages some of which are:
- Shafts which are *not parallel* can also be included in a belt drive system.
- The transmission of torque is possible even at *high speeds*.
- They are elastic drives which help to *dampen speed variations*, vibrations and noise.
- *No lubrication* is required.

Disadvantages are that:
- Some *slip* usually takes place with flat, V-belt or round belt drives.
- The high belt tension results in the *bearings* of the shaft being *heavily loaded.*

7.1.1 Some types of belts and pulleys (Riemen und Scheibenarten)
- **Flat belts** are made out of leather, or out of several layers of leather, canvas and synthetic material.
- **Textile belts** are *endless belts* made of woven polyester or polyamide. They operate with a minimum of vibration and noise and are particularly suitable for driving *internal grinding spindles*.

- **Pulleys for flat belts** (Riemenscheiben für Flachriemen) are made of cast iron, steel, light alloy or plastic. The surface of the pulley has to be *smooth*, otherwise the slip in the belt causes the *wear* to become too *great*. The outer surface of the pulley is *slightly rounded* to ensure that the belt always remains in the centre.
- **V-belts** (Keilriemen) are *endless belts* with a *trapezoidal cross-section*, and are made of rubber reinforced with polyester fibres. In flat belt drives, the friction acts between the inside surface of the belt and the outside surface of the pulley. The friction in V-belt drives acts between the *outer-side surfaces* of the belt, and the *inner-side surfaces* of the pulleys.
- **Ribbed V-belts** (Mehrrippenkeilriemen) *distribute the load* uniformly across the belt. They are used for widely separated axes and heavy loads.

7.1.2 Toothed belt drives (Zahnriemengetriebe)

The transmission of torque in toothed belt drives does not take place through friction, but through the *meshing* of the teeth in the V-belt with the teeth in the pulley. Toothed V-belt drives combine the advantages of flat or ordinary V-belt drives with the *slip-free operation* of the chain drive. Toothed belt drives are characterized by low belt tensioning and a consequent *lowering* of *the load* on the *bearings*. They are particularly useful for shafts which need to have an *exact timing relationship* between each other, as in the case of the crankshaft (or the camshaft) of an internal combustion engine.

7.2 Chain drives (Kettengetriebe)

The transmission of torque and power from one shaft to another can also be accomplished through chain drives. In this type of drive, special toothed wheels called *sprocket wheels* are mounted on the shafts and an endless chain is stretched over them. Some of the advantages of chain drives are:
- High efficiencies of 98 to 99 % even at low speeds.
- No slip takes place, and *no initial tension* is required.
- Chains can move in *either direction.*
- *Heavy loads* of up to 200 kW per chain are possible even at low speeds.
- Chains can link *several shafts* which are spaced wide apart.
- *Idler sprockets* can be used on either side of the chain. These can compensate for slack, guide the chain round obstructions, or change the direction of rotation of the shaft.
- Increased loads can be handled by using *multiple chains*.

Disadvantages are that with small numbers of teeth in the sprocket a *polygon effect* is experienced, which *hinders smooth operation* of the drive. Speed variations, noise and vibrations are the result. This effect can be reduced by having more teeth and using multiple chain drives. The life of a chain is reduced by inadequate lubrication, accumulation of dirt, shocks and vibrations.

8 Bearings (Lager)

Bearings act as **supports** and **guides** for axles and shafts. Bearings are of different types and can be classified according to their function.

1. **Cylindrical journal bearings** (Radiallager) which carry a rotating shaft and have to resist **radial forces**.
2. **Thrust bearings** (Axiallager) which have to resist **axial forces** and prevent longitudinal motion of the shaft.
3. **Guide bearings** (Führungen) which **guide** a **machine element** along the length of a machine **without causing** any **rotatory motion** of the machine element.

8.1 Plain bearings (Gleitlager)

In a plain bearing the shaft rotates in a **shell** or a **bush**. In Fig 5.13 is shown how a **bush** (gudgeon pin) and **split shell bearings** are used in the **connecting rod** of an automobile engine. Frictional forces arise between the shaft and the bearing and these must be kept to a minimum. The frictional forces are minimized by **lubrication**, a term which refers to the process of maintaining a **thin film of oil** between two surfaces which slide past each other. Two types of lubrication are normally used **hydrodynamic** lubrication and **hydrostatic** lubrication.

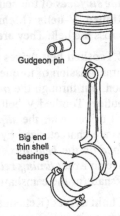

Fig 5.13 Split shell bearings and a gudgeon pin

8.2 The lubrication of plain bearings (Schmierung der Gleitlager)

8.2.1 Hydrodynamic lubrication (Hydrodynamische Schmierung)

In bearings which use the process of hydrodynamic lubrication, the oil film is generated by the **rotational motion** of the **shaft**. When the shaft begins to rotate, the two surfaces are not fully separated by an oil film. As the speed of rotation increases, the oil which is in the **unloaded side** of the bearing will be **forced into the region** where the gap between the surfaces is narrow. The **increasing oil pressure** in the gap moves the shaft, thereby increasing the space between the surfaces. This results in improved lubrication and a consequent reduction of the frictional forces.

8.2.2 Hydrostatic lubrication (Hydrostatische Schmierung)

In bearings which have hydrostatic lubrication, **oil is pumped** into oil pockets in the bearing. The oil flows from these pockets into the **narrow gaps** between the surfaces. The oil pressure in the gap ensures that the shaft and the bearing are **not physically in contact** with each other.

8.2.3 Air lubricated bearings (Gleitlager mit Luft als Schmierstoff)
Lubrication with *air* as the *lubricating medium* (or lubricant) is used in special applications like for example in *measuring instruments*. The frictional forces in such bearings are extremely small.

8.2.4 Other methods of lubrication (Andere Schmierungsarten)
Bearings are provided with oil at low feed rates by such devices as *wicks*, *felt pads* and *drop-feed oilers*. Another method of oiling is *ring oiling*, in which a ring dips into an oil reservoir and rotates round the shaft, smearing it with oil.

8.2.5 Lubrication-free bearings (Wartungsfreie Gleitlager)
Among the bearings that do not need lubrication are:

- Bearings made of *synthetic* and *plastic* materials.
- Sintered material bearings which are *impregnated with lubricant* during manufacture.
- Bearings with special layers which contain a *solid lubricant*.

8.3 Bearings with rolling elements (Wälzlager)
Rolling-contact bearings have a number of rolling elements like *steel balls* or *steel cylinders* which roll between an *outer* and an *inner ring*. *Cages* are used to maintain a *fixed distance* between the rolling elements. The friction is less than in a plain bearing. Particularly advantageous is the low friction at *slow speeds* and when *starting*.

Among the *rolling elements used* are balls, cylindrical rollers, tapered rollers and needle rollers. The rolling elements can be arranged in single or double rows. The rolling elements and the surrounding rings are usually made of hardened steel. Other materials like *stainless steel, ceramics, Monel* (a nickel alloy) and *plastics* are used when *corrosion* has to be prevented. The advantages and disadvantages of bearings with roller elements in comparison with plain bearings are as follows:

Advantages	Disadvantages
1. High load capacity at low speeds.	1. Sensitivity to dirt, contamination.
2. Low friction losses.	2. Sensitivity to shocks.
3. Low lubrication requirements.	3. Limited life and maximum speed.
4. Easy interchangeability due to size standardization.	4. Relatively high level of noise generated.
5. Compensation for the bending of shafts by the use of self-aligning bearings.	5. Smaller load capacity for the same size of bearing.

8.3.1 Ball bearings (Kugellager)
(a) Deep-groove or single row radial bearing (Rillenkugellager)
These are available as *single* (Fig 5.14(a)) or *double row* arrangements, and with single or double row shields or seals. They are used for both *radial loads, and axial* (thrust) loads of (up to 60 % of radial loads). Can be used at *higher speeds*

than is possible with other types of bearings. These are the *cheapest types* of bearings for the same load.

(b) Angular contact bearings (Einreihiges Schrägkugellager)
These are suitable for use with combined radial and thrust loads or heavy thrust loads in a single direction (Fig (5.14(b)). Should be *used in pairs* mounted as *mirror images* of each other. The pairs are referred to as *duplex bearings*.

(c) Double row angular contact bearings (Zweireihiges Schrägkugellager)
In this bearing, a pair of angular contact bearings are mounted in such a way that they are *mirror images* of each other (Fig 5.14(c)). This may be considered to be a *composite duplex bearing*.

(d) Internal self-aligning double row bearings (Pendelkugellager)
These bearings can be used for radial and axial (thrust) loads (Fig 5.14(d)). These are *self-aligning bearings* which can withstand slight axial and angular *misalignments*.

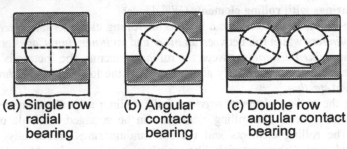

| (a) Single row radial bearing | (b) Angular contact bearing | (c) Double row angular contact bearing |

Fig 5.14 Different types of ball bearings

8.3.2 Roller bearings (Rollenlager)
(a) Cylindrical roller bearings (Zylinderrollenlager)
These bearings (Fig 5.15(a)) can be used with *heavy radial* and *light thrust* loads. They can be used with large diameter shafts and have the highest speed limits for roller bearings.

(b) Tapered roller bearings (Kegelrollenlager)
These bearings (Fig 5.15(b)) can be used with *heavy radial and thrust* loads. They are only used in pairs arranged as *mirror images* of each other. They are also available in double row form as *duplex bearings*.

(c) Spherical roller bearings (Pendelrollenlager)
These bearings (Fig 5.15(c)) have a *self-aligning capability* which enables them to compensate for any *bending* or *misalignment* of the shaft. They are excellent for heavy radial and moderate thrust loads.

(d) Needle bearings (Nadellager)
The rolling elements in these bearings (Fig 5.15(d)) are *long in comparison* to their *diameter* and are very useful where *space is a factor*. They can be used with or without an *inner race*. If used without an inner race, the *shaft* has to be *hardened* and *ground*. They *cannot be used* for *axial* (thrust) *loads*.

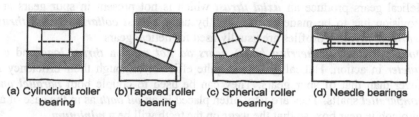

| (a) Cylindrical roller bearing | (b)Tapered roller bearing | (c) Spherical roller bearing | (d) Needle bearings |

Fig 5.15 Different types of roller bearings

8.3.3 Thrust bearings
(a) Axial thrust ball bearings (Axial-Rillenkugellager)
These can only used with *axial loads* and with shafts which are perpendicular to the bearing. A ball bearing for unidirectional loads is shown in Fig 5.16 (a).

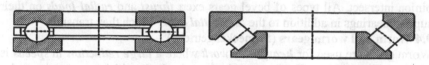

Fig 5.16 (a) Axial thrust ball bearing   (b) Axial thrust roller bearing

(b) Axial thrust roller bearings (Axial-Pendelrollenlager)
These bearings shown in Fig 5.16(b) are *self-aligning* and can compensate for *small shaft misalignments*.

9. Gears (Zahnräder)
The coupling of two shafts which are rotating at the same speed or at different speeds can be achieved in many ways. One way is by the use of gears or (gear wheels), which are wheels with *teeth cut in them*. The ratio of the speeds of rotation of the shafts is dependent on the *ratio of the number of teeth* in the wheels. Both *parallel* and *nonparallel* shafts can be coupled by gear wheels.

9.1 Spur gears (Stirnräder mit Geradverzahnung)
Spur gears are the most commonly used types of gears and have teeth cut *parallel* to the gear's *axis of rotation*. They are used to couple shafts which are parallel to each other. The *smaller wheel* is called the *pinion* and the *larger wheel the gear*. The teeth have *involute* profiles. Cycloidal teeth are used in precision mechanical clocks, but not in gears used for the *transmission of power*. Spur gears can have *internal* and *external* teeth as shown in Fig 5.17 (a) and (b).

9.2 Helical gears (Stirnräder mit Schrägverzahnung)
The teeth of a helical gear lie on a cylinder and are cut at an *angle* to the *axis of rotation* of the gear (Fig 5.18(b)). The teeth on helical gears *mesh* with each other *progressively*, and are therefore *smoother* and *quieter* in action than the teeth on spur gears. A further advantage is that the *life* of the gears is *longer* for the same loading than that of equivalent spur gears.

Helical gears produce an *axial thrust* which is not present in spur gears and provision has to be made to take this by using *thrust collars* or *axial thrust bearings*. Involute profiles are usually used for helical gears.

Double helical (or *herring bone*) *gears do not exert* a *thrust load* and are *quieter* in action. Helical gears are quite efficient although their efficiency is lower than that of spur gears. They can be used to couple both *parallel and nonparallel* shafts. They are very often placed in an *oil bath* as in the case of an automobile gear box, so that the *wear* on the teeth will be a *minimum*.

9.3 Bevel gears (Kegelräder)

Bevel gears are most often used to couple shafts which are at *right angles* to *each other* (Fig 5.18(a)). The teeth of these gears do not have an involute profile. The teeth lie on a *conical surface* and the apex of the pinion and the mating gear intersect at the point at which the axes of the gear shaft and the pinion intersect. All types of bevel gears exert *thrust* and *radial loads* on their support bearings in addition to the *tangential loads* which they transmit.

9.4 Worm and worm gears (Schneckengetriebe)

Worm gears are used for *heavy duty work* where a *large reduction* in *speeds* is required (Fig 5.18(c)). They are used to couple shafts which are at right angles to each other. The *driving wheel* is called the *worm* and the driven wheel the *worm gear*. The worm has a form which is similar to the screw thread of a cylindrical bolt. The screw thread meshes with the teeth on the worm gear (wheel). The teeth on the worm wheel are *cut at an angle* to its axis of rotation.

9.5 Rack and pinion (Zahnstangengetriebe)

In this device, a *flat piece of metal* with *teeth cut on it* and called the *rack*, meshes with a *spur gear* called the *pinion* (Fig 5.17 (c)). *Rotation of the pinion* causes the rack to move in a *straight line*. This is an example of the conversion of rotatory motion into linear motion. The *drill feed mechanism* of a drilling machine uses a rack and pinion, as does the *coarse feed mechanism* of an optical microscope.

9.6 Gear boxes (Zahnrad-Stufengetriebe)

It is quite often necessary to *repeatedly change* the *ratio of the speeds* of rotation of two shafts called the *gear ratio*. Such changes are done in *several stages* through several pairs of shafts and gears. A gear assembly which does this smoothly and efficiently is called a *gear box* (Fig (5.18(d)). Gear boxes are an integral part of machine tools and automobile engines. It must be noted that only a few *selected gear ratios* can be obtained in this way. Intermediate values of gear ratio cannot be obtained.

9.7 Continuously variable speed drives (Stufenlose Getriebe)

Drives in which the ratio of the *output speed* to the *input speed* can be *varied continuously* cannot be constructed using gear wheels. However other types of continuously variable speed drives are available. Among these are friction drives, hydrostatic drives and electrical drives.

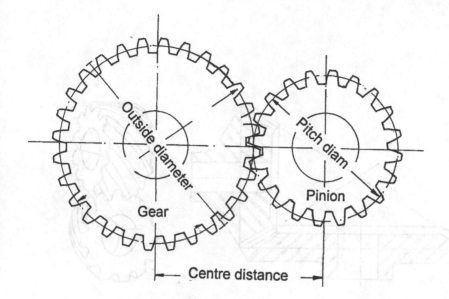

Fig 5.17(a) Spur gears

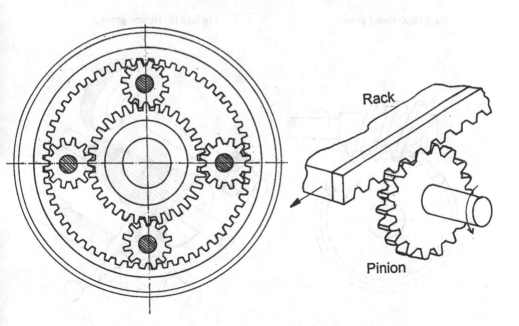

Fig 5.17(b) Involute epicyclic gearing   Fig 5.17(c) Rack and pinion drive

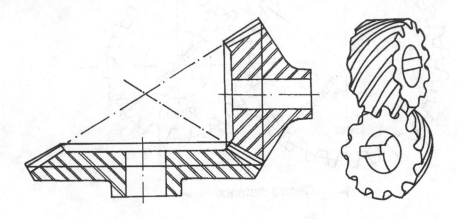

Fig 5.18(a) Bevel gears   Fig 5.18 (b) Helical gears

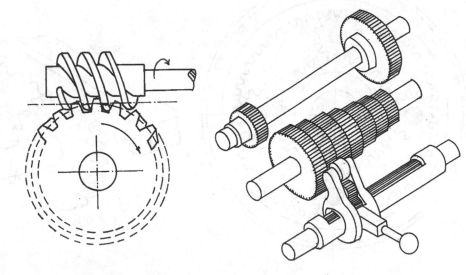

Fig 5.18 (c) Worm and worm gear   Fig 5.18 (d) Gear box fitted to a lathe

VI Joining processes (Verbindungsarten)

1 Adhesive bonding (Klebverbindung)

Materials which *cannot be soldered* or *welded* can be bonded by the use of adhesives. This is usually better than using metal fasteners. Adhesives are used to bond *metals to nonmetallic materials* like plastics, glass, wood, porcelain, etc., or where welding is *detrimental* to the properties of the material. It is particularly useful with very *thin materials*, which cannot be joined by welding or with the help of metal fasteners. Adhesive bonding offers great advantages in aircraft construction, because by using the method of *sandwich construction*, very light components with *excellent stiffness properties* can be produced.

1.1 Adhesives (Klebstoffe)

The adhesives used are either phenol resins or epoxy resins. Sometimes rubber dissolved in a solvent is also used.

1.1.1 Contact adhesives (Kontaktklebstoffe)

These adhesives based on *rubber solutions* are applied both surfaces. The surfaces are exposed to air for a short time and then pressed together.

1.1.2 Resin based adhesives (Klebstoffe auf Kunstharzbasis)

These are usually composed of *two components*, a binder and a catalyser. The two are mixed in equal parts just before use. The mixture is first applied to the two surfaces which have to be joined, after which the two surfaces are pressed together. It takes some time for the adhesive to harden, and the hardening time can be shortened by increasing the temperature, or by adding an accelerator as a third component to the mixture.

Cold-hardening adhesives harden at room temperature, while warm-hardening adhesives harden only at high temperatures. There are also slow-hardening adhesives as well as fast-hardening adhesives, which can both be used at room temperature.

1.2 Preparation of surfaces for bonding (Vorbehandlung der Oberflächen)

An adhesive bond can only be strong and effective when the surfaces to be bonded are *clean* and *free of oil and fat*. Removal of dirt, oxide and paint remnants can be done by *chemical* or *mechanical methods*. Chemical methods have been said to produce better bonding in aluminium, magnesium and copper alloys as also with glass and ceramics. Removal of fat, oil and wax is done by using chemicals like perchloroethylene, methylchloride or acetone. Etching particularly on metal surfaces is done by using dilute sulphuric acid.

1.3 The bonding process (Klebverbindungsprozess)

It is important to spread the adhesive evenly over the surface. When using a rubber solution as an adhesive, the *time* at which the surfaces are brought together under pressure after exposure to air *is important* for the strength of the bond. With the other adhesives only one surface needs to be smeared with adhesive. The surfaces can be brought together *immediately* thereafter.

2 Soldering (Löten)

Soldering is a process which is used to join two pieces of metal by a third metal alloy called solder. The solder should have a *lower melting point* than the objects being soldered. Some of the requirements and features of the process are the following:

1. The solder *is melted* during the soldering process, but the metals being joined are *not melted* unlike in welding, where localized melting takes place.

2. A solder must have the ability to *wet the surfaces* being joined, which means that it is able to form an *alloy* with the metals. At the same time, the solder must melt at a temperature well below the melting points of the metals which are being soldered. Alloys of *tin and lead* satisfy these requirements, because tin alloys readily with iron, with the copper alloys and with lead. At the same time, tin-lead alloys melt at temperatures between 183°C and 250°C, well below the melting points of the metals to be joined.

3. A soft soldered joint *lacks mechanical strength*. It is desirable that the objects to be joined be *fastened* to each other *securely* by flanging, folding, twisting, spot welding, etc., to give the joint sufficient mechanical strength. Soldering makes a joint *gas-tight* and *water-tight*. It also ensures *good electrical contact*.

4. When two metal objects are joined to each other before soldering, there is usually a *gap between the surfaces* into which the solder must flow. Only then are the capillary forces large enough to make the solder flow into the gap.

5. The *working temperature* of a solder is the temperature at which the solder wets, flows, and alloys with the metal surface. The solder must be in *liquid form* and not in the form of a paste. A good soldered joint is only possible if those parts of the metal objects to be soldered and the solder are *all heated* to the *working temperature* of the solder during the soldering process.

6. Before soldering, it is necessary to *clean the surfaces* of the objects to be soldered to remove fat, oxides and other substances like remaining paint, etc. This is first done by scrubbing or other *mechanical methods* if necessary. Further cleaning is done by *coating* the surfaces *with a flux* before soldering. These are substances which clean the surfaces, and are of two types, the first being *acidic flux* which is *corrosive*, and the second being *resin* which is *noncorrosive*. The fluxes used in the workshop are strongly acidic in character. Commonly used types are hydrochloric acid, zinc chloride and acid paste. For electrical work a *noncorrosive flux* like *resin* is required. The solder is usually in the form of a wire having a core of resin.

2.1 Soft soldering (Weichlöten)

The working temperature for soft soldering is below 450°C. The parts to be soldered and the solder are *brought into contact* and *heated* to the required temperature. Heating may be carried out in many ways. A *soldering iron* is used in sheet metal work and also in electrical work. The tip is made of copper and is heated by electricity or gas. The *copper tip* conducts the heat quickly to the *soldering point* which is heated together with the solder. The solder flows into the joint and solidifies when the iron is removed. Other ways of heating are by using a gas flame, by using an oven, by floating in a bath of molten solder, by induction heating, etc. The solder may be fed into a joint by hand or placed on the joint before soldering in the form of washers, rings, sheets, shims,etc.

3 Brazing (Hartlöten)

Brazing is preferred to soldering when a *tougher, stronger joint* is required. This is a versatile process and can be used to *join virtually all metals*. The temperatures used for brazing are above 450°C, but below the melting temperatures of the metals to be joined. Here again cleanliness of the surfaces is essential, and the fluxes used are of the *borax or fluoride* type.

The types of solder used are (1) *brazing solder (or spelt)* which are copper-zinc alloys (of the brass type) and (2) *silver solder* which is an alloy of silver, copper and zinc. Silver solder has lower working temperatures of between 610°C and 850°C.

For joining steel components, the brass (copper-zinc) type brazing solder is used, while silver solder is used for metals like copper, brass and silver. The use of a *gaseous atmosphere* to protect the components being brazed from oxidation is often necessary. There are many ways of carrying out the brazing process. Among these are dip brazing, furnace brazing, resistance brazing, etc.

4 Welding (Schweißen)

Welding is the term applied to a process of the joining of metals, by heating to temperatures high enough to cause *localized melting* of the metals at the joining points. This process may be carried out with or without the use of *filler metals,* and with or without the use of *pressure*. This is different from soldering or brazing, where any melting of the base metals does not takes place. Welding produces *strong joints* which are almost as strong as the base metal, except in cases of *fatigue loading*.

Many types of joints are possible. Among these are butt joints, tee joints (fillet joints), lap joints, corner joints, etc. Examples of these are shown in Fig 6.1. There are many types of welding processes and it is possible to group them in many ways. Here they have been divided into two groups (1) Fusion welding processes and (2) pressure welding processes.

4.1 Fusion welding (Schmelz-Schweißen)

In fusion welding processes, metals are joined by melting them locally at the joining points *without* using *pressure*. A filler metal may or may not be used.

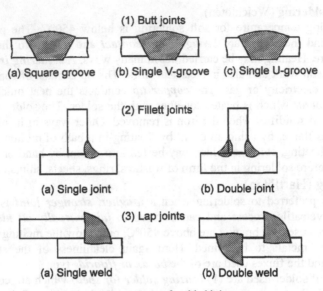

Fig 6.1 Some types of welded joints

4.1.1 Oxy-acetylene welding (Gasschmelzschweißen)

In oxy-acetylene welding, a mixture of *oxygen* and *acetylene* is burned in a *blowpipe*. The temperature can be controlled varying the amounts of by gases flowing through the pipe. The flame produced can exceed a temperature of 3000°C. This is well *over the melting points* of the metals being welded. The regions of the *metals at the joint melt,* and mix with each other. The joint is made stronger by the melting of a filler rod which is held close to the joint. A small pool of metal is formed at the joint, and this pool solidifies when the flame is removed forming a strong joint. The process is illustrated in Fig 6.2.

4.1.2 Electric arc welding (Metall-Lichtbogenschweißen)

In electric arc welding, an arc is maintained between the metals to be welded and an electrode, or between two electrodes. Both *direct* and *alternating* currents may be used. An alternating current welding installation uses a *transformer* to supply the welding current. The transformer changes the high voltage low current input from the mains supply, to a low voltage high current source at the output terminals which are used for welding. The current flows through the *filler rod* which also acts as an *electrode*.

The arc is produced between the tip of the filler electrode and the metal. Here again the temperature can be above 3000°C, and small portions of the metal and the filler rod are melted. Movement of the electrode causes the deposition of a *pool of metal* along the path of the electrode. The type of set-up used is shown in Fig 6.3.

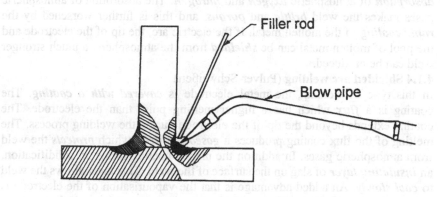

Fig 6.2 Oxy-acetylene welding

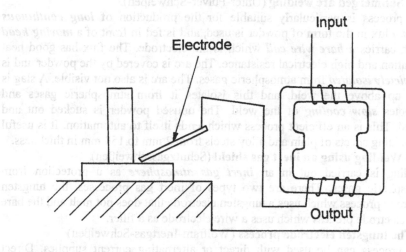

6.3 Electric arc welding

4.1.3 Disadvantages of bare metal electrodes
(Nachteile der ungeschützten Elektroden)

The welds produced by arc welding are considerably weakened by the *absorption* of atmospheric *oxygen* and *nitrogen*. The absorption of atmospheric gases makes the weld *brittle* and *porous*, and this is further worsened by the *rapid cooling* of the molten metal. If the electric arc, the tip of the electrode and the pool of molten metal can be *shielded* from the atmosphere, a much stronger weld can be produced.

4.1.4 Shielded arc welding (Pulver-Schweißen)

In this type of welding, the metal electrode is *covered with a coating*. The coating is a *flux* which has a higher melting point than the electrode. The coating extends beyond the tip of the electrode during the welding process. The melting of the flux coating produces a *gaseous shield,* which *protects* the weld from atmospheric gases. In addition the melted coating forms on solidification, an *insulating layer* of slag on the surface of the weld. This layer allows the weld to *cool slowly*. An added advantage is that the vapourisation of the electrode is reduced by the shielding effect, which results in a more economical use of the welding electrode.

Welds produced by this process are *strong, reliable* and *ductile*. The automatic welding of steel parts is possible by this process. Other applications are in the welding of mild and alloy steels, stainless steels, and to a lesser extent in the welding of nonferrous metals.

4.1.5 Submerged arc welding (Unter-Pulver-Schweißen)

This process is particularly suitable for the production of *long continuous welds.* Flux in the form of powder is used, and is fed in front of a *moving head* which carries a *bare wire coil* which is the electrode. The flux has good heat insulation and high electrical resistance. The arc is covered by the powder and is *completely isolated* from atmospheric gases. The arc is also not visible. A slag is built up above the weld, and this isolates it from atmospheric gases and promotes *slow cooling* of the weld. The unused powder is sucked out and reused. This is an efficient process which lends itself to automation. It is useful for welding sheets of plain and alloy steels from 2 mm to 150 mm in thickness.

4.1.6 Welding using an inert gas shield (Schutzgasschweißen)

Welding is carried out in an *inert gas atmosphere* as a protection from atmospheric gases. There are two types of inert gas processes, the tungsten electrode process which uses a tungsten electrode that does not melt and the bare wire electrode process which uses a wire electrode as a filler.

(a) The tungsten electrode process (Wolfram-Inertgas-Schweißen)

This process can be used with direct or alternating current supplies. Direct current is used to weld alloy steels and also nonferrous metals and their alloys. The *bare tungsten electrode* is surrounded by a tube through which *argon gas* flows. The argon gas forms an inert shield which gives protection from the atmosphere. A filler rod is used to improve the weld. As *no flux* is used, *no slag* is formed resulting in a clean weld.

(b) The wire electrode process (Metall-Schutzgasschweißen)

In this process, an arc is formed between the metal parts being welded and a *metal wire electrode* which also acts as a filler. The electrode melts and this adds to the pool of metal formed by the melting of the base metals. Argon gas *flows through a tube* which surrounds the wire forming a *protective shield*. The wire is fed from a reel and the feeding rate is automatically controlled, so that the gap of the arc between the tip of the wire and the parts being welded is kept constant. The whole unit is mounted on a *moving head* which travels automatically along the path of the weld. Long continuous runs can be achieved using this process. This method is particularly useful for the production of clean, slag-free welds in components made from stainless steel and nonferrous metals.

4.1.7 Electron beam welding (Elektronenstrahlschweißen)

In this process a *filament* of a *metal* like *tungsten* is heated to emit electrons. The electrons arc *accelerated* towards an anode which has a *high positive voltage* relative to the filament. The number of electrons reaching the anode can be controlled by a grid whose potential can be varied. The electrons pass through a hole in the anode and then through a focusing coil and various deflecting coils before reaching the workpiece. On striking the workpiece, the *kinetic energy* of the electrons is *transformed* into *heat* which melts the portions of the metal to be welded. The set-up has to be operated in a *high vacuum* to prevent oxidation of the filament.

The extremely *high energy density* of the electron beam is effective in producing *narrow deep welds* at high speed with minimum distortion. The welds made are said to be stronger than those made by other welding processes. The welds are also *very pure,* because they are made in a vacuum.

Disadvantages are the difficulty of *aligning the beam* with the joint to be welded, and the difficulty of manipulating the filler metal in a vacuum. The cost, complexity, and maintenance difficulties are also problems.

4.1.8 Laser beam welding (Laserschweißen)

A laser is an optical source which generates a monochromatic *beam of coherent light* which usually has a very small cross-section. This *highly concentrated* beam can be used as an energy source for welding. The beam is concentrated through concave mirrors or lenses to the location where the welding has to take place.

A vacuum is not required for a laser welding operation, and this process has many of the advantages of electron beam welding at a *lower cost* and at *higher rates of production*. High speed production of narrow deep wells is possible and complex welding operations using *computer control* are possible. Lasers can also be used for cutting and scribing of metals and nonconductive materials like ceramics. The main disadvantage is that the *power output* of lasers is *limited*.

4.2 Pressure resistance welding processes (Widerstandspreßschweißen)

All the processes mentioned so far produce fusion welds *without* the application of *pressure*. In each case, a liquid pool of metal was formed between the two parts to be joined, with the further addition of more liquid metal by the use of a filler rod or an electrode. The *process is continued* along the joint and on cooling the *liquid metal solidifies* and the two parts are welded together.

In pressure resistance welding, an *electric current* is made *to flow through* a *circuit* of which the *workpiece is a part*. The resistance of the circuit is a maximum at the interface of the parts being joined, and the *heat generated* is sufficient to cause a local fusion of the metal. The parts to be welded are *pressed together* at the same time, and a welded joint is produced without the use of a filler material.

4.2.1 Spot welding (Punktschweißen)

In spot welding, metal sheets are held together under pressure and joined permanently by welding them at *individual spots*. The size and shape of the welds are *controlled* by the size and shape of the electrodes which are usually circular. This method is usually used for welding thin metal sheets. The speed of welding can be improved by using machines with *multiple electrodes*. This is a *cheap, rapid* and *efficient* technique and each spot weld takes only a few seconds to carry out. A typical set-up for spot welding is shown in Fig 6.4.

4.2.2 Seam welding (Rollennahtschweißen)

Seam welding is used when *air-tight* or *water-tight* seams are required. This is achieved by making a succession of spot welds in such a way that the *welds overlap*. The process is illustrated in Fig 6.5. The welding current is applied *intermittently* by using timing devices while the sheets are pressed together by revolving circular electrodes. The process is *continuous* and results in welds, each weld overlapping its neighbours. *Stitch welds* can also be made using the same equipment. Stitch welds are a series of welds *separated* by a *constant distance* as shown in Fig 6.5.

4.2.3 Butt or upset welding (Abbrennstumpfschweißen)

This process is used to weld pieces of metal having approximately the *same cross-section*. A good example is in the construction of a *lathe tool*. There is no reason why the entire tool should be made of high speed steel, and a cheaper but equally good tool can be made by welding two different types of steel. The cutting end of the steel is made of *high speed steel* and butt welded to a shank (or rear end) made of medium *carbon steel*.

Butt welding is achieved by bringing together the two parts to be joined, and passing the *welding current* through the *joint*. When the joint reaches the desired temperature, *further pressure* is applied and the welding current is switched off. This technique can only be used with small areas of contact where a good match between the surfaces exists. *Special preparation* of the surfaces may be necessary.

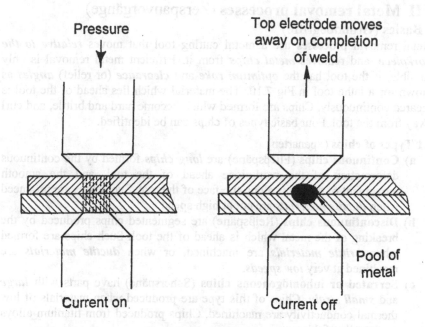

Fig 6.4 Spot welding

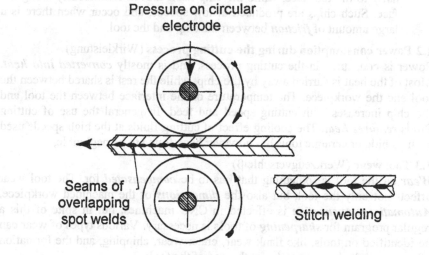

Fig 6.5 Seam and stitch welding

VII Metal removal processes (Zerspanvorgänge)

1 Basics (Grundbegriffe)

Metal removal processes use a metal cutting tool that moves *relative to the workpiece,* and removes *metal chips* from it. Efficient metal removal is only possible, if the tool has the *optimum rake* and *clearance* (or relief) *angles* as shown for a lathe tool in Fig 7.10. The material which lies ahead of the tool is sheared continuously. Chips are formed which become hard and brittle, and curl away from the tool. Four basic types of chips can be identified.

1.1 Types of chips (Spanarten)

   a) **Continuous chips** (Fließspäne) are *long chips* formed by the continuous deformation of the workpiece ahead of the tool, and the smooth movement of the chips along the face of the tool. Such chips are produced when *ductile materials* are cut at high speed.
   b) **Discontinuous chips** (Reißspäne) are segmented chips produced by the breaking of the metal which is ahead of the tool. Such chips are formed when *brittle materials* are machined, or when *ductile materials* are machined at very *low speeds*.
   c) **Serrated or inhomogeneous chips** (Scherspäne) have parts with *large* and *small strain*. Chips of this type are produced when materials of low thermal conductivity are machined. Chips produced from titanium alloys are usually of this type.
   d) **Built-up edge chips** (Scheinspäne) arise when a piece of *metal attaches itself* to the tool face, while the chip itself moves continuously along the face. Such chips are produced at low speeds, and occur when there is a large amount of *friction* between the chip and the tool.

1.2 Power consumption during the cutting process (Wirkleistung)

Power is consumed in the cutting process, and is mostly *converted into heat*. Most of the heat is carried away by the chip, while the rest is shared between the tool and the workpiece. The temperature of the interface between the tool and the chip increases with cutting speed and feed. In general the use of cutting fluids *removes heat*. The cooling effect of cutting fluids at the high speeds used with carbide or ceramic tools has however been found to be negligible,

1.3 Tool wear (Werkzeugverschleiß)

Wear on the tool is something that has to be *compensated* for. The tool wear affects not only the tool, but also the *dimensions* of the machined workpiece. *Automatic compensation* is effected in CNC machines, but in spite of this a regular program for *sharpening* of tools is necessary. Various types of wear can be identified on tools, like flank wear, crater wear, chipping, and the formation of grooves called *wear notches* at the end of the tools.

When metals are cut at high speeds, long spirals of chips can form and become *entangled* with the tooling. *Chip breakers* are often introduced on the tool to prevent this from happening. These cause the chips to break into small sections.

1.4 Materials for cutting tools (Schneidstoffe)

a) **Carbon steels** (Kohlenstoffstahl) are used for low speed applications. They are *inexpensive*, but have largely been replaced by better types of tool steels.

b) **High speed steels** (Schnellarbeitsstahl) are the *most used* alloy tool steels. They have the ability to maintain their strength, hardness, and cutting edge over long periods of time. This has led to their use in a variety of cutting, drilling, reaming, broaching and other tools.

c) **Cemented carbide** (Hartmetalle) tools are made from metal carbides by the use of *powder metallurgy* techniques. They have a high level of hardness, high thermal conductivity, and low thermal expansivity. Their elastic modulus is high, and they are not subject to plastic flow. They are used in the form of *small tips* which are brazed or mechanically fastened to a steel shank. Typical of the carbides used are tungsten carbide with cobalt as a binder, or titanium carbide with nickel and molybdenum as binder.

d) **Ceramic or oxide** (Schneidkeramik und Oxidkeramik) **inserts** are *harder* than cemented carbides. Ceramic materials *maintain their hardness* and wear resistance at temperatures of up to 1200°C. They are however brittle and sensitive to variations in cutting forces. They are therefore used without cooling, under conditions where cutting forces remain constant. They can be used at higher cutting speeds than cemented carbide tools, but are *unsuitable* for use with *aluminium alloys*. Oxides used are mainly grains of *corundum* (aluminium oxide) which are bonded together by the use of powder metallurgy techniques.

e) **Polycrystalline diamond tools** (Polykristalliner Diamant (PKD)) are used where high *dimensional accuracy* and a *good surface finish* are required. These tools are made by bonding a layer of polycrystalline diamond on a carbide substrate. They are particularly helpful for use with *nonferrous metals* which are not easy to machine with normal steel tools.

f) **Polycrystalline cubic boron nitride (CBN)** (Polykristallines Bornitrid) (PKB)) is the *next hardest material* to diamond. Polycrystalline CBN is bonded to a carbide substrate before being used in cutting tools. It is useful in machining high temperature alloys and ferrous alloys. Diamond and CBN are also used as abrasives for grinding.

1.5 Cutting fluids (Kühlschmierstoffe)

Cutting fluids which consist of liquids or gases are allowed to flow over the tool and workpiece to ensure *smooth operation* of the cutting proceses. Cutting fluids have *many functions*. Among them are, keeping the work and tool cool, ensuring lubrication which reduces power consumption and tool wear, providing a good surface finish on the work, helping in the satisfactory formation of chips and their removal, and preventing corrosion.

In most cases, the fluid is pumped from a sump and allowed to *flow over* the *cutting interface*. Mist cooling is also used. In this type of cooling, water-based fluids are dispersed as fine droplets in air.

Some of the main types of cutting fluids are the following:

a) **Air drafts** (Luftströme) are used mainly to remove dust or small chips during grinding, polishing and boring operations. A certain amount of cooling is also obtained.

b) **Emulsions** (Wassermischbare Kühlschmierstoffe) are produced by emulsifying a soluble oil with water. The ratio of the oil to water is between 1 to 10 and 1 to 100. Emulsions *contain additives*, which *reduce friction* and provide *effective lubrication* at the interface between tool and chip in a machining operation. These are low cost cutting fluids, which are used for practically all types of cutting and grinding operations.

c) **Lubricating oils** (Mineralölhaltige Kühlschmierstoffe) are also used in metal cutting operations. They are used in cases where *lubrication* is *more important* than cooling. The oils usually used are mineral oils with varying proportions of fat, sulphur and chlorine.

d) **Solutions** (Mineralölfreie Lösungen) are cutting fluids which contain various *chemical agents* such as amines, chlorine, nitrites, nitrates, phosphates etc, in water. Most of these fluids are coolants although some are lubricants.

2 The use of hand tools for metal removal
(Spannende Formgebung von Hand)

Most of the metal components and objects manufactured today are mass produced using machines. However it is nearly always necessary to use hand tools *for repairs*, and also for the *creation of models* and *prototypes*.

2.1 Marking-out process (Vorbereitung durch Anreißen)

Before work can be started on a piece of metal, it is usually subjected to a marking-out process. In this process, suitable lines are scribed on the surface of the metal, to *help* and *guide* the person who is working on the metal. Some of the marking out tools are shown in Fig 7.2 (b).

The work is carried out on a very flat and level table which is made of cast iron or hard stone. The lines are drawn on the surface using a hard scriber. Scribing blocks, vernier height gauges with scriber inserts and other devices are used in this process. Centre punches are used to mark points, and dividers are used to scribe circles.

2.2 Holding and clamping devices (Spannelemente)

When working on a metal by hand, it is necessary to use devices which hold the metal workpiece firmly in place. A vice *holds an object firmly* between its jaws while work is being done on it. Other holding devices are toolmaker's clamps, vee-blocks and angle blocks (See Fig 7.2 (a)).

Cylindrical rods and pipes are held in *vee-blocks* while work is being done on them. *Angle blocks* have two faces at *right angles* to each other, and are used when the position of the work has to be changed by 90°.

2.3 Measuring devices (Meßgeräte)

Measuring devices are required during the marking-out process, and also for checking the dimensions of the work when it is finished. Vernier calipers can be used to measure internal, external and depth dimensions to an accuracy of 1/100 cm. Micrometers are *more accurate* than vernier calipers, and can be used to make similar measurements. Inside and outside calipers can be used to check the dimensions of an object while *machining* is being done on it. Try squares are used to check if two surfaces are *perpendicular* to each other (Fig 7.2 (a) & (b)).

2.4 Hand tools (Handwerkszeuge)

Chisels are of many types and are used for *removing metal* from an object as well as for shearing metal into two or more parts. Chisels can also used for work on wood and other materials.

Saws are used to cut metals as well as other materials. Their construction depends on the materials on which they are to be used. Hand saws are of many types and include wood saws for cutting wood, hack saws for cutting metal, and slitting saws for cutting slots in screw heads. Machine driven saws like hack saws, band saws and circular metal saws save *time* and *labour*.

Files of various shapes are used for *metal removal*, and available types range from coarse to fine. *Scrapers* can be used to remove very small amounts of metal when a very *flat surface* is required.

Hammers are used for many purposes like shaping metal, hammering nails, driving a chisel, etc. A hammer with a *head* made of wood or other *soft material* is called a *mallet*.

Punches are of various types. A *centre punch* is used to mark a specific point on a piece of metal. A *pin punch* is used to drive-in or remove pins and keys. *Drift punches* are used to align two or more pieces of metal which are to be joined together by bolts or rivets.

Taps are used for cutting *internal screw threads*, and dies are used for cutting *external screw threads*. A tap and a tap wrench which holds the tap when it is being turned to cut the thread are shown in Fig 7.1(b). Also shown are a circular split die and a die stock (or holder) for the die.

2.5. Screw drivers, spanners and keys (Schraubenzieher und Schlüssel)

Screw drivers are used to turn screws, which have a slit cut in their head. *Offset screw drivers* are used for turning screws in awkward places. *Pliers* are used for holding, gripping, turning, etc. A *spanner* or a *wrench* is a tool used for turning nuts and bolts. There are many types of spanners like the ring spanner, socket spanner, double-ended spanner, etc. *Keys* are special tools used for turning screws. They can be inserted into *slots* which have been cut into the head of a screw, after which the screw can be rotated. A screw with a *hexagonal slot* in its head can be turned with a *hexagonal key*.

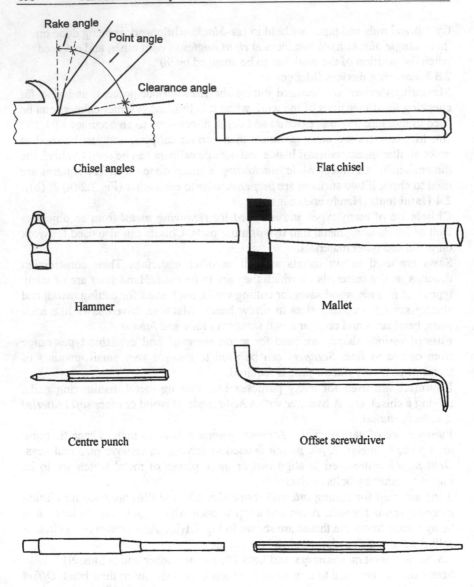

Rake angle

Point angle

Clearance angle

Chisel angles

Flat chisel

Hammer

Mallet

Centre punch

Offset screwdriver

Drift punch

Pin punch

Fig 7.1(a) Some hand tools

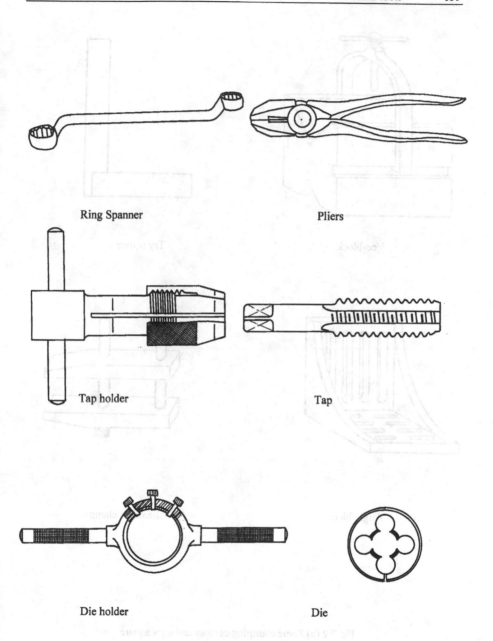

Ring Spanner

Pliers

Tap holder

Tap

Die holder

Die

Fig 7.1 (b) Some hand tools

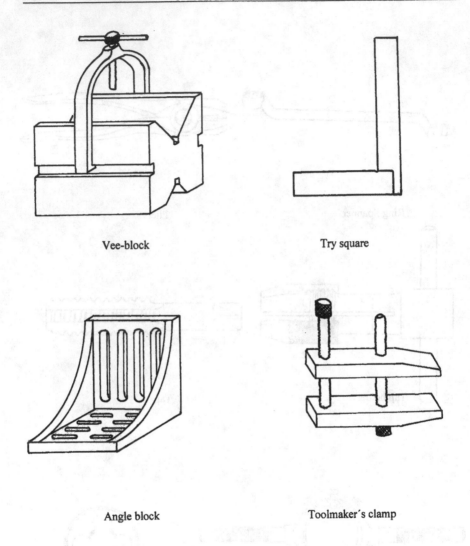

Vee-block   Try square

Angle block   Toolmaker´s clamp

Fig 7.2 (a) Some clamping devices and a try square

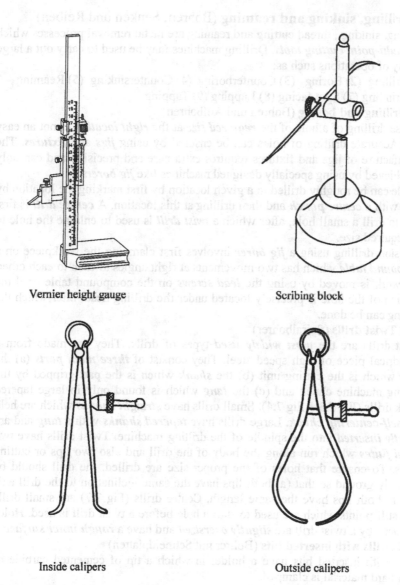

Vernier height gauge   Scribing block

Inside calipers   Outside calipers

Fig 7.2 (b) Some marking-out and measuring equipment

3 Drilling, sinking and reaming (Bohren, Senken und Reiben)

Drilling, sinking , thread cutting and reaming are metal removal processes which use *multi-point cutting tools*. Drilling machines may be used to carry out a large variety of operations such as:

(1) Drilling (2) Boring   (3) Counterboring (4) Countersinking (5) Reaming (6) Grinding (7) Spot facing (8) Lapping (9) Tapping

3.1 Drilling and boring (Bohren und Aufbohren)

Precise drilling of a hole of the *required size* at the *right location* is not an easy task. Accurate drilling of holes can be ensured by using *jigs* and *fixtures*. The manufacture of jigs and fixtures requires extra care and precision and can only be achieved by using specially designed machines like *jig borers*.

A hole can be roughly drilled in a given location by first marking the location by hand with a *centre punch* and then drilling at this location. A *centre drill* is first used to drill a small hole, after which a *twist drill* is used to enlarge the hole to the required size.

Precision drilling using a *jig borer* involves first clamping the workpiece on a *compound table* which has two movements at right angles to each to each other. The work is moved by using the *lead screws* on the compound table until the position of the hole is precisely located under the drilling head, after which the drilling can be done.

3.1.1 Twist drills (Spiralbohrer)

Twist drills are the *most widely used* types of drills. They are made from a cylindrical piece of high speed steel. They consist of *three main parts* (a) the *body* which is the cutting unit (b) the *shank* which is the part gripped by the drilling machine chuck and (c) the *tang* which is found only in large tapered shank drills (Fig 7.3 &Fig 7.4). Small drills have *straight shanks* which are held in a *self-centering chuck*. Large drills have *tapered shanks* with a *tang* and are *directly inserted* into the spindle of the drilling machine. Twist drills have two *spiral flutes* which run along the body of the drill and also two lips or cutting edges. To ensure that holes of the proper size are drilled, the drill should be precisely ground so that (a) both lips have the same inclination to the drill axis and (b) both lips have the same length. Centre drills (Fig 7.7) are small drills with stiff points which are used to start a hole before a twist drill is used. Holes produced by a twist drill are *slightly oversized* and have a *rough inner surface*.

3.1.2 Drills with inserted bits (Bohrer mit Schneidplatten)

Drills with inserted bits have a holder in which a tip of cemented carbide or other hard material is clamped.

3.1.3 Boring bars (Bohrstangen)

Boring bars have *inserted tips* of hard material. The diameter of the hole that can be bored is *adjustable*.

3.2 Reaming (Reiben) Reaming is a *finishing process* used to ensure that a hole has the *right size*. The hole is drilled slightly undersize and then enlarged to the correct size by using a reamer which removes only a *small amount of metal*.

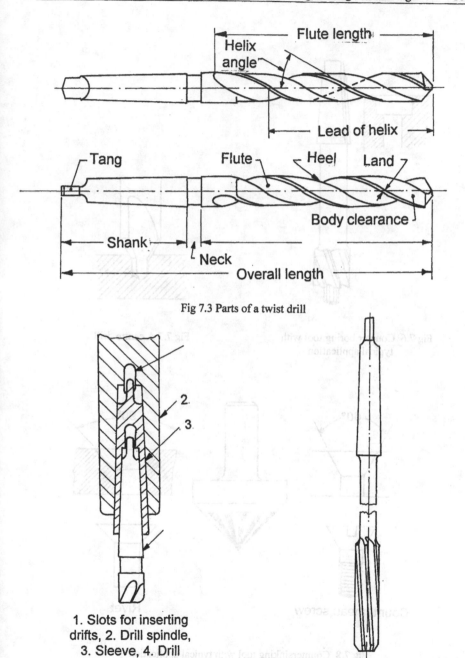

Fig 7.3 Parts of a twist drill

1. Slots for inserting
drifts, 2. Drill spindle,
3. Sleeve, 4. Drill

Fig 7.4 Drill, sleeve and spindle

Fig 7.5 One type of machine reamer

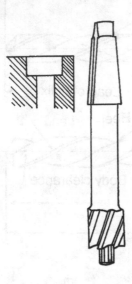

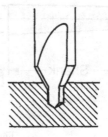

Fig 7.6 Counter boring tool with
typical application

Fig 7.7 A centre drill

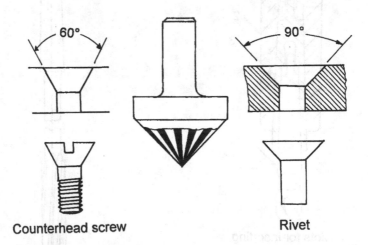

60°

Counterhead screw

90°

Rivet

Fig 7.8 Countersinking tool with typical application

3.3 Counterboring and countersinking (Zylindersenken und Kegelsenken)

Counterboring is usually used to *enlarge* the top of a hole cylindrically to accommodate the *heads of bolts*. The tool used is called a *counterbore* (Fig 7.6). Deeper counterboring can also be carried out for other purposes. When a hole has to have different diameters at different depths, special *multidiameter* drills can be used. These have different diameters at different lengths of the drill, and ensure that the different diameters are concentric.

Countersinking enlarges the top of a hole *conically* to accommodate the *head* of a *countersunk screw* or rivet. The tool used is called a *countersink* (Fig 7.8).

3.4 Spot facing (Planansenken)

This is an operation which *levels the surface* around a hole that has been drilled or counterbored to take the head of a screw.

3.5 Tapping (Gewindebohren)

This is the operation of *cutting internal threads* by means of a cutting tool called a *tap* (Fig 7.1(b)). The tap has threads and flutes cut on it and has cutting edges which are hardened and ground. A hole of the right size has to be drilled before the tap is used. If the tap is now screwed into the hole, it cuts the required internal thread. *Exeternal threads* are cut by using *dies* (Fig 7.1(b)).

3.6 Drilling machines (Bohrmaschinen)

Drilling machines usually have a column, a table, and a drilling head. The workpiece is placed on the table. Both the drilling head and the table can be moved up and down to accommodate different sizes of workpieces. The *speed* of rotation of the drill and the *rate of downward feed* can be adjusted, as can the *depth of hole* to be drilled.

3.6.1 Radial drilling machines (Schwenkbohrmaschinen) have in addition a

radial arm on which the drilling head is mounted. The radial arm can be *rotated* to any position on the column, and in addition the head may *move sideways* on the radial arm. The movements enable a hole to be drilled at any position on a large workpiece. In some machines the drill head can be swung about a horizontal axis perpendicular to the arm. This enables the drilling of holes at an *angle to the vertical*.

3.6.2 Multiple drilling machines (Mehrspindelbohrmaschinen)

These are machines with *several spindles* which can drill several holes in a workpiece simultaneously. They may be used to drill the *same pattern* of holes in a large number of workpieces as part of a mass production program. Jigs may be used to guide the drills.

3.7 Clamps, jigs and fixtures (Spannelemente und Vorrichtungen)

It is necessary to hold the work securely while drilling, and various types of clamps are used for this purpose. *Fixtures* are used to ensure that holes are drilled in the *right locations,* and *jigs* are used when the drills *have also to be guided* into the holes.

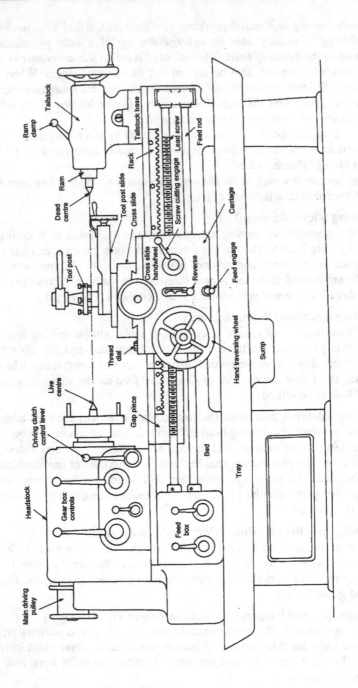

Fig 7.9 Shows the different parts of a typical lathe

4 The lathe and single point cutting tools
(Drehmaschine und Drehmeißel)

The lathe is the most important of all machine tools, and is used for producing components which are *symmetrical* about *an axis*. It can be used for machining cylindrical external and internal surfaces, and also for the turning of *conical* and *tapered* surfaces. In addition a lathe can be used to cut *screw threads* on an already machined cylindrical surface.

The lathe is a versatile machine which is widely used in tool rooms to perform a wide variety of work. The accuracy of the work done on a lathe depends on the *skill* and *experience* of the operator. A lot of time is spent on tool setting, tool changing etc., with the result that it is *unsuitable* for use in *production work*. It is mainly used for the making of prototypes and spare parts. A typical lathe is shown in Fig 7.9.

4.1 The size of a lathe (Baugröße der Drehmaschine)
The *specifications* for the size of a lathe include the following items:
1. **The height of centres** (Spitzenhöhe) measured above the lathe bed.
2. **The length between centres** (Spitzenweite) which corresponds to the maximum length of work that can be mounted between lathe centres.
3. **The maximum bar diameter** (Spindelbohrdurchmesser). This is the maximum diameter of bar stock (metal bars) that will pass through the hole in the headstock spindle.

4.2 Parts of a lathe (Bauteile der Drehmaschine)
A lathe has a rigid bed with *parallel guideways* on which are mounted a fixed *headstock* and a movable *tailstock*. In addition there is a *carriage* which can be moved along the guideways of the bed, in a direction which is parallel to the *axis of rotation* of the spindle. The headstock has a strong spindle which is driven by a motor through a *gearbox*.

The *speed* of the *spindle* can be varied through a wide range, to suit the type of work that has to be done. A lathe is also usually fitted with a *lead screw* which is used for *screw-cutting*. The lead screw can be geared to the headstock spindle through the gear box.

Knurling is another operation that can be carried out on a lathe. This is done by using a knurling tool which consists of a set of hardened steel rollers mounted in a holder. The rollers have a definite pattern of teeth cut on them. This creates a diamond (or other) shaped pattern on the surface of the workpiece.

4.3 Lathe accessories (Drehmaschinen Zubehör)
Lathe accessories are used for *supporting* the work or *holding* the tool. They include lathe centres, catch plates and carriers, chucks, collet chucks (or colletts), face plates, angle plates, mandrels, steady and follower rests.

4.3.1 Lathe centres (Zentrierspitze)
The workpiece is held very often between a *live centre* (Mitlaufende Zentrierspitze) on the spindle, and a *dead centre* (Einfache Zentrierspitze) on the tailstock. A *half centre* allows the facing of the ends of a bar without removing

the centre. A *rotating* or frictionless centre can be fitted in the tailstock for heavy work revolving at high speed. This type of centre is fitted with roller and thrust bearings.

4.3.2 Carriers and catch plates (Mitnehmer und Mitnehmerscheibe)

A workpiece which is held between centres is usually driven by using carriers (or driving dogs) and catch plates. Catch plates are attached to the end of the headstock spindle and the carriers attached to the end of the workpiece by set screws.

4.3.3 Chucks (Futter)

Chucks are some of the most important devices for holding a workpiece in a lathe. The chuck is usually *screwed-on* to the *nose* of the *lathe spindle*. The following are some of the most important types of chucks that are available:

1. **Three jaw self-centering** or **universal chuck** (Dreibackenfutter) – This chuck is useful for holding round, hexagonal or other *regular shaped* workpieces. The work is *centered automatically* (although the centering may not be very accurate), because all three jaws move forward or backward by an equal amount when the chuck is adjusted.

2. **Four jaw chuck** (Vierbackenfutter) – Here each jaw may be *adjusted independently* and this chuck is particularly suitable for heavy and *irregular shaped* workpieces.

3. **Magnetic chuck** (Magnetfutter) – These chucks are used for holding very *thin workpieces* which cannot be held in an ordinary chuck. They can also be used when the distortion caused by the jaws of an ordinary chuck is not desirable.

4. **Collets or collet chucks** (Spannzange) – These are small chucks that fit into the *headstock spindle* and are used for holding *bar stock* (long bars or rods). These chucks are particularly useful in cases where accurate centering and quick setting are required.

5. **Face plates** (Planscheiben) – Face plates are circular plates which can be fitted by screw threads to the nose of the lathe spindle. Face plates have *slots* on them for *holding work* by *bolts and clamps*.

6. **Mandrels** (Drehdorne und Spanndorne) – A mandrel is used for holding and rotating a *workpiece* that *has a bore* in it. The mandrel is mounted between centres and the workpiece rotates with it. Many types of mandrel are available.

4.3.4 Steady and follower rests (Feststehende und mitlaufende Setzstocke)

Steady and follower rests are used when turning *long bars* or other similar workpieces. These rests *prevent the bending* of the bars which would be caused by the cutting forces. The steady rest is *bolted to the bed* of the lathe, while the follower rest is *bolted to the saddle* of the lathe. Care is required when adjusting the pads of the rests. Excessive pressure is not required, and the workpiece should be able to rotate with reasonable ease.

4.3.5 Attachments for lathes (Zusatzgeräte)

Attachments are *additional devices* used for *specific purposes*. They include stops, thread chasing dials, and attachments for taper turning, milling, grinding, gear cutting, knurling, etc.

4.4 Single point lathe cutting tools (Drehmeißel)

The lathe uses a single point cutting tool for the removal of metal from a workpiece. The action is *similar* to that of *a chisel*. The pointed part of the tool is a wedge which presses on the metal and *tears off a chip* when the metal moves relative to the tool. Efficient metal cutting is only possible if the cutting tools are made of the right material and have the *correct cutting angles* ground on the tip of the tool. A lathe cutting tool with the main cutting angles is shown in Fig 7.10.

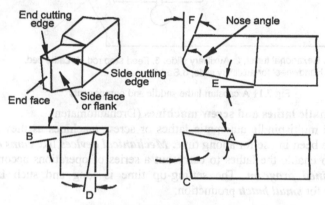

A:Top or front rake angle, B:Side rake angle
C:Front clearance or relief angle, D:Side clearance or relief angle
E:Side cutting angle, F:End cutting angle

Fig 7.10. Typical lathe tool showing the main cutting angles

4.5 Capstan and turret lathes (Revolverdrehmaschinen)

The tool room lathe is unsuitable for production work, but *modified versions* of the lathe like the capstan and turret lathes have been used as *mass production machines*. These machines have the same headstock and four way tool post as the ordinary lathe. However the tailstock is replaced by a *hexagonal turret* and each face of the turret can carry one or more tools. These tools may be used successively to perform a *series of different operations* in a *regular sequence*. The feed movements of each tool may be *regulated by stops*. Lead screws are not fitted and the cutting of screw threads is done by using taps and dies. The *initial setting* of tools is a *skilled operation*. Once this has been done a *semi-skilled operator* can produce a large number of components in a short time.

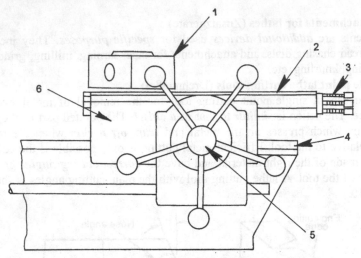

1. Hexagonal turret, 2. Auxilliary slide, 3. Feed stop rod, 4. Lathe bed,
5.Handwheel for auxilliary slide, 6.Saddle

Fig 7.11 A capstan lathe saddle and components

4.6 Automatic lathes and screw machines (Drehautomaten)

Single and multispindle automatic lathes or screw machines as they are often
called have been in use for a long time. *Mechanical devices* like *cams and stops*
are used to enable the lathes to carry out a series of operations according to a
predetermined program. The setting-up time is long and such lathes are
unsuitable for *small batch* production.

5 Milling machines (Fräsmaschinen)

A milling machine uses a *multipoint tool* to remove metal from a workpiece.
The use of multipoint tools enables the machine to achieve *fast rates* of *metal
removal* and produce a *good surface finish*.

5.1 Column and knee type of milling machines (Konsolfräsmaschinen)

These machines have a column and a projecting knee which carries the saddle
and work table. The work table has *three independent movements*, longitudinal,
transverse and vertical. Milling machines *lack the rigidity* required for heavy
production work and are mainly used in tool rooms and workshops.

5.2 The horizontal milling machine (Horizontal Fräsmaschine)

A horizontal milling machine (Fig 7.12) has a *horizontal spindle* located in the
upper part of the column. It receives power from a motor through belts, gears
and clutches. The spindle *projects slightly* out of the column face and has a
tapered hole into which *cutting tools* and *arbors* may be inserted. An arbor is an
extension of the machine spindle. The *overhanging arm* which is fixed on top of
the column serves as a *bearing support* for the arbor. The arbor has a *taper
shank* which fits into the nose of the machine spindle.

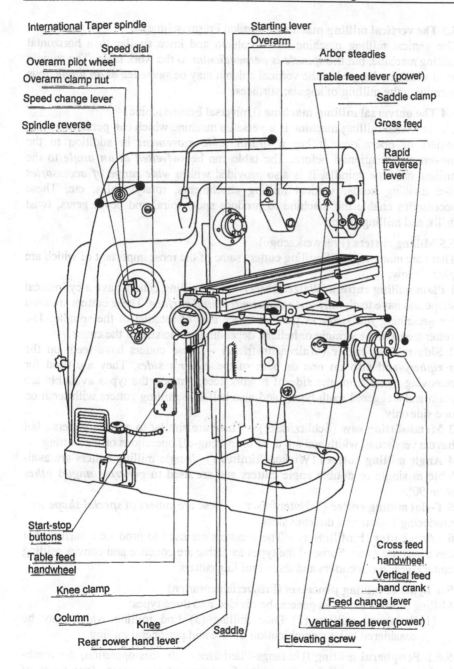

International Taper spindle
Speed dial
Overarm pilot wheel
Overarm clamp nut
Speed change lever
Spindle reverse

Starting lever
Overarm
Arbor steadies
Table feed lever (power)
Saddle clamp
Gross feed
Rapid traverse lever

Start-stop buttons
Table feed handwheel
Knee clamp
Column
Knee
Rear power hand lever
Saddle

Cross feed handwheel
Vertical feed hand crank
Feed change lever
Vertical feed lever (power)
Elevating screw

Fig 7.12 A horizontal milling machine

5.3 The vertical milling machine (Vertikal Fräsmaschine)
The vertical milling machine has a column and knee similar to a horizontal milling machine, but the *spindle is perpendicular* to the work table. The spindle head which is clamped to the vertical column may be *swivelled at an angle*, thus permitting the milling of angular surfaces.

5.4 The universal milling machine (Universal Fräsmaschine)
The universal milling machine is a versatile machine which can perform a wide variety of operations. It has a *fourth table movement* in addition to the movements mentioned before. The table can be *swivelled at an angle* to the milling machine spindle. It is also provided with a *wide range of accessories* like dividing heads, vertical milling attachments, rotary tables, etc. These accessories enable the machine to produce spur, spiral and bevel gears, twist drills, and milling cutters.

5.5 Milling cutters (Fräswerkzeuge)
There are many types of milling cutters some of the most important of which are given below.
1 Plain milling cutters (Walzenfräser) – Plain milling cutters have a cylindrical shape and have teeth only on the *circumferential surface*. These cutters are used for generating *flat surfaces* parallel to the axis of rotation of the spindle. The cutter teeth can be straight or helical, depending on the size of the cutter.
2 Side milling cutter (Walzenstirnfräser) – These cutters have teeth on the *periphery* and also on one or both of the *cutter's sides*. They are used for removing metal from the side of a workpiece. Among the types available are plain and staggered teeth types, and also half side milling cutters with teeth on one side only.
3 Metal slitting saw (Schlitzfräser) – These are similar to milling cutters , but have a very small width and are used for parting-off operations or for slotting.
4 Angle milling cutters (Winkel-Stirnfräser) – Angle milling cutters are available as single or double angle cutters and are used to *generate angles other than* 90°.
5 T-slot milling cutter (T-Nutenfräser) – These are cutters of *special shape* for producing T-slots and dovetail slots.
6 Form cutter (Profilfräser) – These cutters are used to produce a surface that has a *definite form*. Some of the types available are concave and convex milling cutters, gear tooth cutters and thread milling cutters.

5.6 Types of milling processes (Fräsverfahrensarten)
Milling processes may in general be divided into three types:
  (1) Peripheral milling (2) Face milling (3) End milling which may be considered to be a combination of face and peripheral milling.

5.6.1 Peripheral milling (Umfangs-Planfräsen) – In this operation, the machined surface is *parallel* to the *axis of rotation* of the cutter. Two types of processes are possible, depending on the *sense of rotation* of the cutter relative to the *direction of movement* of the workpiece. These processes are termed *up milling* and *down milling*.

(a) Up milling (Gegenlauffräsen) is the *conventional type of milling* in which the cutter is rotated *against the direction* of travel of the workpiece. The thickness of the chip is small at the beginning and increases towards the end of the cut. The upward cutting force tends to *lift the workpiece* from the table. Due to the nature of the cutting forces, difficulty is experienced in feeding coolant at the beginning of the cut. In spite of the disadvantages, it is used so often because it is *safer*. It is particularly useful for milling castings containing particles of sand, and also for milling welded joints.

(b) Down milling (Gleichlauffräsen) which is also called climb milling, is a process in which the cutter is rotated in the *same direction* as the *direction of travel* of the workpiece. The thickness of the chip and the cutting force are a maximum when the tooth begins cutting, and a minimum when the tooth stops cutting. In this case, the *clamping* of the work is *easier* and the chips are disposed more easily. Coolants can be fed into the cutting zones, and this reduces heat problems and gives a *better surface finish*. However down milling cannot be used on machines having *backlash*. The backlash causes the work to be drawn below the cutter at the beginning of the cut, and leaves the work free when the cut is over. This results in vibration and damage to the workpiece.

5.6.2 Face milling (Stirn-Planfräsen) produces a flat machined surface *perpendicular* to the *axis of rotation* of the cutter. The *main cutting* is done by the *peripheral cutting teeth* while the face cutting edges finish the work by removing a small quantity of metal.

5.6.3 End milling (Stirn-Umfangs-Planfräsen) – This may be considered to be a combination of peripheral and face milling. The milling cutter has teeth on the periphery as well as on the end face.

6 Broaching (Räumen)

Broaching is a metal removal process in which a single tool having a *definite geometric shape* is used. The tool which has multiple cutting edges is *pushed* or *pulled* along the inner or outer surfaces of a workpiece. Only a small amount of metal can be removed in this process, and it is necessary most of the metal should have been removed previously by other machining processes. Broaching is carried out when special profiles having a good surface finish, and good dimensional accuracy are required. The work is carried out in a *single operation* and high production rates with *unskilled labour* can be achieved.

6.1.1 Internal broaching (Innenräumen) is used to produce holes of *various shapes* in cylindrical holes which have been previously made by drilling, boring, casting, etc. For example *keyways* or *splines* may be cut by a broach in a previously drilled hole. It is also possible to cut *internal spirals* in a workpiece, by using special broaches and by rotating the workpiece during the broaching process. This is similar to the cutting of screw threads by using taps.

6.1.2 External broaching (Außenräumen) can be used to generate external surfaces with profiles similar to those produced by other machining procedures.

7 Surface finishing processes (Oberflächenfeinbearbeitung)

Machined surfaces are usually rough, and may not meet the standards of *surface quality* and *dimensional accuracy* required. Finishing operations are used to bring the surfaces of the components to the required standard. Very little metal is removed in a finishing operation, and previous machining operations should have been carried out satisfactorily before the finishing processes are started.

7.1 Grinding (Schleifen)

Grinding is an operation that is carried out by using a rotating abrasive wheel to remove metal from an object. Very high levels of dimensional accuracy and surface finish can be achieved by grinding. Comparatively little metal is removed usually 0.25 to 0.50 mm in most operations.

Grinding can also be used effectively to machine materials which are too hard to be machined in any other way. Some of the grinding processes which are normally in use are the following:

(1) External grinding including *centreless grinding* (2) Internal cylindrical grinding (3) Surface grinding (4) Form grinding

7.1.1 External cylindrical grinding (Außen-Rundschleifen) is used to grind a cylindrical or tapered surface on the outside of the workpiece. The workpiece is rotated between centres as it is moved lengthwise, while making contact with a rotating grinding wheel which rotates at high speed.

7.1.2 Internal cylindrical grinding (Innen-Rundschleifen) is carried out in a similar way to external grinding by using smaller grinding wheels that grind the inner surfaces of the bored workpiece. The small wheels need to rotate at high speeds for effective grinding.

7.1.3 Centreless grinding (Spitzenlosschleifen)

This is a method of external grinding used to grind cylindrical, tapered and formed surfaces that are not held between centres. The set-up consists of a work-rest which lies between a grinding wheel and a regulating (or back-up) wheel. The work is placed on the work-rest, which moves forward with the regulating wheel thus forcing the work against the grinding wheel.

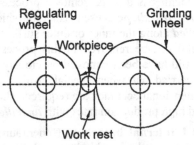

Fig 7.13 Centreless grinding

7.1.4 Surface grinding (Planschleifen) is an operation used to grind flat surfaces. The grinding is carried out by using either the *periphery* or the *end face* of the grinding wheel. The workpiece is given a reciprocating movement below or on the end face of the grinding wheel.

7.1.5 Form grinding (Profilschleifen) uses grinding wheels which are *specially shaped* to *accurately finish* surfaces which have been previously machined to a special shape like gear teeth, spline shafts and screw threads.

7.1.6 Abrasive wheels (Schleifkörper)

Grinding wheels contain *very small abrasive particles* of a *very hard material* like silicon carbide or aluminium oxide. The wheels are made by using a *bonding material* which holds the abrasive particles together. The particles have sharp edges and each wheel acts like a *multi-tooth cutter* which removes metal from the workpiece. Different particle sizes and different kinds of bonding materials are used to make a *whole range* of grinding wheels.

Silicon carbide wheels are useful for grinding materials of *low tensile strength* such as cutting tool tips, ceramics, cast iron, brass, etc. Aluminium oxide wheels are better suited for materials of *higher tensile strength* such as most kinds of steel, wrought iron, tough bronzes, etc. Diamond wheels which are made by *impregnating* a metal wheel with *fine diamond particles* are used for special purposes like gem cutting.

7.2 Lapping (Lappen)

Lapping is a process which is used to produce *geometrically true surfaces*, achieve high dimensional accuracy, secure a fine finish and obtain a close fit between two surfaces. *Very little material* is removed (less than .01 mm), and this method cannot correct for substantial *errors in form*.

The lapping process uses a lapping paste and a tool called a lap. The paste is formed by mixing fine particles of an abrasive material with oil. The lap is made of a relatively *soft porous material* like cast iron or copper. The paste is rubbed into the lap, causing the *abrasive particles* to become *imbedded* in it. In the lapping process, the work is rubbed with the lap in an *ever changing path*. Laps can be operated by machine or by hand.

7.3 Honing (Honen)

Honing is a grinding process which is mostly used for *finishing cylindrical holes* by means of *bonded abrasive stones* called hones. Honing can be used to remove as much as 3 mm of material, but is normally used for removing less than 0.25 mm. Honing is primarily used to correct errors in roundness, taper, axial distortion, and tool marks in a previously machined workpiece.

The abrasive tool is in the form of a *flat stone* or *stick* called a hone. A few of these stones are mounted *round a metal cylinder* to form a honing tool. This tool is reciprocated axially while being in contact with the rotating workpiece. Coolants are used to remove small chips and and keep temperatures uniform.

7.4 Superfinishing (Kurzhubhonen)

Superfinishing is a *honing process* which uses large bonded abrasive stones to produce a surface of *extremely high quality*. A *very thin layer* of metal (less than .02 mm) is removed in this process. It may be applied to the internal or external surfaces of objects made of cast iron, steel or nonferrous alloys which have been previously ground or machined.

In this process a stick containing a *very fine abrasive* is placed in a suitable holder and applied to the surface of the workpiece with light pressure. The abrasive holder is given an oscillating motion, while the workpiece is rotated or given a reciprocating motion depending on the shape of the surface. A *special lubricant,* usually a mixture of kerosene and oil is used to obtain a high quality finish. Superfinishing is used for many types of components such as crankshaft bearings, cylinder bores, pistons, valve stems, and other moving metallic parts.

7.5 Polishing and buffing (Polieren und Hochglanzpolieren)

Polishing is an operation which is used to remove scratches, pits, tool marks and other *defects* from *rough surfaces*. In polishing the *dimensional accuracy* of the polished surface is *not important*. Polishing wheels are made of leather, canvas, felt or wool. The abrasive particles are glued to the surface of the wheel and the wheel is rotated while the object to be polished is *held against it* until the desired finish is obtained.

Buffing produces a *much more lustrous surface* than is obtainable by polishing. Buffing wheels are made of felt, leather or pressed and glued layers of cloth. The abrasive is *mixed with a binder* and is applied to the buffing wheel or to the work. The wheels are rotated with a *high peripheral speed* of up to 40 m/s. The abrasives used are iron oxide, chromium oxide, emery, etc. The binder is a paste consisting of wax mixed with grease, paraffin, turpentine, etc.

7.6 Shot or grit blasting (Körnchenblasen)

Shot or grit blasting is a *surface cleaning process* in which abrasive or other particles moving at high speed are *made to strike the surface* to be cleaned. Rust, scales, burrs, etc, are removed and the surface acquires a *matt appearance*.

7.7 Shot peening (Verfestigungsstrahlung)

Shot peening is a process used to *strengthen and harden* a surface. In this process, *steel balls* moving at high speed *strike the surface* which becomes *work hardened* and *fatigue resistant.*

7.8 Barrel finishing (Trommelpolieren)

This process *eliminates hand finishing* and is therefore very *economical* in the use of labour. The workpieces are placed in a many-sided barrel together with abrasive materials (like stones, abrasives, etc) and a suitable liquid. When the barrel is rotated for an appropriate amount of time, the *mutual impact* between the workpieces and the abrasive materials *removes surface irregularities*.

VIII CNC machines (CNC Maschinen)

1 Introduction (Einführung)

1.1 Development of CNC machines (Entwicklung der CNC Machinen)

The machining of metal components had been done in the past by using *conventional machine tools* like lathes, milling machines, etc. and the accuracy of the work done was dependent on the *skill and experience* of the operator. A lot of time was taken for *tool setting, tool changing*, etc. The result was that these machines were unsuitable for large scale production.

An improvement was made by using machines like capstan and turret lathes. These had *rotating turrets* for rapid tool changing and *adjustable stops* which allowed feed control. However the setting time was long, and these machines were unsuitable for *small batch production.*

A further improvement was made by using single and multispindle *automatic lathes.* These were able to produce large numbers of identical components rapidly. The operations were carried according to a *predetermined program* and were controlled by *mechanical devices* like *cams* and *stops.* However, the setting time was long, and these were also unsuitable for small batch production.

A solution to this problem was found by developing machines that were able to produce components *automatically* in accordance with a series of predetermined instructions called a program. The use of a *programmed computer* to control a machine tool, led to this form of automation being called *computer numerical control* (CNC).

1.2 Features of CNC machines (Besonderheiten der CNC Maschinen)

CNC machines are capable of doing more than producing components automatically. They possess the characteristics of *extreme accuracy, repeatability, reliability and high productivity.* No *setting or resetting* of these machines is required. All that one needs to do when a new component is required is to write out a *new program*, which when used correctly, produces a component that has precisely the dimensions called for in the program.

The program contains the *control instructions* which guide the machine, and the *geometric data* required to produce the component. Extensive changes in the structure and operation of existing machine tools had to be carried out before the new CNC machines could be built. Some of the changes made are given below.

1. CNC machines have *heavier structures* with improved rigidity and better *damping* characteristics. This gives them the ability to withstand *large cutting forces* and ignore the *thermal effects* of the chips produced. The machines are often placed in a *tilted position* thus allowing the chips to slide down.
2. Bearings and spindles have *zero play.* All spindles, drive elements and systems are *dynamically balanced.*

3. Recirculating ball screw systems with *low friction* and *zero backlash* are used. They have a mechanical efficiency of over 98 %.
4. A *continuous indication* of the *position* of the cutting tool on each axis is produced and transmitted back to the axis drive, which continues to drive the tool until the *target position* is reached.
5. Each tool is mounted in a *holder* (or adapter) and is *preset* to the correct length. Tool length *compensation* is made for actual differences between preset and actual tool length after the tool has been reground or changed.
6. A large number of tools are usually required and these are *stored* in disc, ring, chain or other types of *magazine*.
7. Automatic tool changing systems enable a *direct exchange* of tools from the magazine into the machine spindle. This done by using *grippers* in an *automatic tool change cycle*.
8. Automatic *workpiece changing systems* are used to change workpieces. The workpieces are clamped to *pallets* which can be *accurately located* and clamped to the work holding surfaces of the machine. Pallett changing installations carry the work to the machine and back again.
9. The main spindle and axis feed drives have servo (or other) motors with *continuously variable speed* and *feed rates. Load fluctuations* are *compensated* for to avoid speed variations. All motions are *independent* of *opposing forces* due to cutting loads, friction or inertia.

1.3 Method of operation of a CNC machine
(Arbeitsweise einer CNC Maschine)

A CNC machine does the same kind of work that a skilled operator does, but it does it without *human intervention.* The work is done *automatically* and with more *precision* than the work done by an operator. The method of operation of a CNC machine may be summarized as follows:

1. The machine is controlled by a *program* which is recorded on a paper tape, magnetic tape or other data carrier. Information which controls the relative motion between tool and workpiece, as well as geometric data relating to the component is contained on the tape in *digital form*. The tape is fed into the system to start the process.
2. A processing unit converts the data into *electrical signals* that the machine tool can understand.
3. The data is stored in a *memory unit* until it is required for use.
4. The stored data is converted into *machine movements* by *servo units* on the machine tool.
5. A measuring system *measures* the machine *movements* and feeds back the measured values to a comparison unit. This unit compares measured and target values and instructs the drive unit to continue the movement until the target value is reached. In this way a *closed loop feedback* control system ensures that *target values* are reached (Fig 8.1).

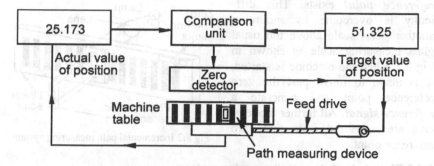

Fig 8.1 Closed loop feedback control system for positioning

1.4 Path measuring systems (Wegmeßsysteme)

The accuracy of a CNC machine is *entirely dependent* on the *accuracy* of its path measuring system. Path measuring systems can be of different types depending on the measuring process whether it is direct or indirect, absolute or incremental, analog or digital.

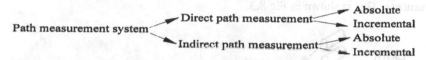

1.4.1 Direct path measurement (Direkte Wegmessung)

In direct path measurement, measurements are made without the use of any *intermediate mechanical device*. The linear movement of a milling machine table can for example be measured *directly* by using an *optical scale*. The accuracy of the measurement will depend on the accuracy of the scale.

1.4.2 Indirect path measurement (Indirekte Wegmessung)

In indirect path measurement, the measurements are carried out by using an intermediate mechanical device. Mechanical devices like ball screws or rack and pinions, need a *rotary encoder* or a *resolver* to generate an electrical signal.

1.4.3. Digital measuring systems (Digitale Meßsysteme)

Digital path measuring systems are usually based on *optoelectrical principles*. A ruled scale is usually attached to the table and the movement of the scale can be measured by using a *detector*. The smallest resolution of the scale is 5µm. However path increments of 1µm can be measured by using an *electronic interpolator*.

1.4.4 Digital incremental path measuring systems
(Digital-inkrementale Wegmeßsysteme)

In a digital incremental path measuring system, any point at which the carriage is in when the *system is started* is taken as the *zero point*. However, the programming of distances along the axes is only possible provided that a *fixed reference point* exists. This difficulty is overcome by mounting another glass scale above the usual glass measuring scale as shown in Fig 8.2. When the machine is started, it is made to move past the zero reference point and generate a *reference signal*. All further movements are measured from this zero reference point.

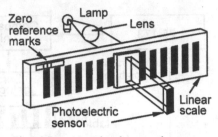

Fig 8.2 Incremental path measuring system

1.4.5 Digital absolute path measuring systems
(Digital-absolute Wegmeßsysteme)

In this type of measurement, all distances are measured from a *special reference point*. A special glass scale is used with successive columns of squares which are either black or transparent. Each column has five squares where the squares are arranged vertically to correspond to successive binary digits. The columns are so arranged horizontally that they correspond to successive decimal numbers. This is shown in Fig 8.3.

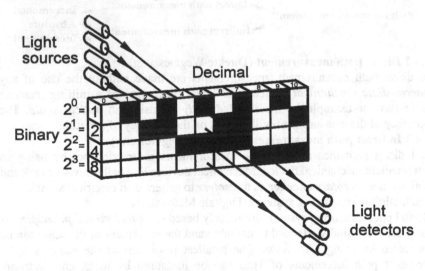

Fig 8.3 Digital absolute path measuring systems

2 Geometrical basis for programming (Geometrische Grundlagen)

2.1 Coordinate system (Koordinatensystem)

A coordinate system with *rectangular axes* is used to describe the motion of tools and workpieces. It is based on the "right hand rule", involving the thumb, middle and index fingers of the right hand. The thumb indicates the direction of the X axis, the index finger the Y axis, and the middle finger the Z axis.

2.1.1 Choice of axes for machine tools (Anordnung der Koordinatenachsen)

Z axis – The direction of the *main spindle* axis is taken to be the Z axis.
X axis – The X axis is the larger of the two other axes and if possible horizontal.
Y axis – The Y axis is *perpendicular* to the ZX plane.

2.1.2 Secondary axes (Zusätzliche Achsen)

If extra axes are required parallel to the X,Y,Z axes, these are designated U,V,W, with U parallel to X, V parallel to Y and W parallel to Z.

2.1.3 Rotational axes (Drehungen um die Koordinatenachsen)

The designation of a rotational axis is based on the *linear axis* about which rotation occurs. A refers to rotation about the X axis, B refers to rotation about the Y axis, and C refers to rotation about the Z axis.

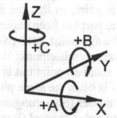

Fig 8.4 Rotation about the axes

2.2 Zero points and reference points (Nullpunkte und Bezugspunkte)

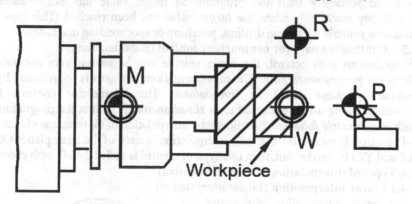

Fig 8.5 Shows some of the zero and reference points

2.2.1 Machine zero point M (Machinennullpunkt M)

The machine zero point is at the origin of the machine's coordinate measuring system. This is *fixed* and cannot be moved. In most cases it lies on the *spindle axis* of the machine. In practice the machine zero point should be automatically reached, when the machine is *switched on.* It should also be possible to reach zero by pressing a particular key, or by using an instruction in the program itself.

2.2.2 Machine reference point R (Referenzpunkt R)

With some CNC machines, it is not possible to reach the machine zero point. In such cases, another *more convenient point* is adopted as the machine's reference point. Instead of starting measurements along the axis from zero, measurements are now made from the *coordinates* of the *new reference point.*

2.2.3 Workpiece (or piece part) zero point W (Werkstücknullpunkt W)

The workpiece zero point can be *freely selected* by the programmer. However on a lathe or other turning machine, it usually chosen to be at the intersection of the spindle axis with the *left* or the *right* edge of the workpiece.

2.2.4 Program zero point P (Programmnullpunkt P)

Coordinate values taken with reference to the machine zero point are not suitable for use in programming. A program zero point is usually chosen with reference to the *piece part* or the *tool change point*. All geometric data are related to this point, so that the machining process can start and continue soon after the piece part has been clamped on the machine.

2.2.5 Tool change point TCP (Not standardized)
 (Werkzeugwechselpunkt WWP – nicht genormt)

This is not a standard point but is very often the *starting point* for the machining process. This is often the same point as the program zero point.

2.3 Control modes (Steuerungsarten)

2.3.1 Point to point control (Punktsteuerung)

In this method of control, *rapid traverse* (or rapid movement) takes place on all axes independently until the programmed target value has been reached. *Machining starts* only when the *target value* has been reached. This type of control is suitable for use on drilling, punching or spot welding machines.

2.3.2 Continuous path (or contouring) control (Bahnsteuerung)

In continuous path control, the cutter spindle can be *moved* very accurately along any *programmed path*. The movement along the axes is coordinated by a *software package* called an *interpolator*. The interpolator controls the movement along all programmed axes *simultaneously*, so that the programmed path is *accurately followed*. Simultaneous interpolation of all three axes is called 3-D control. If successive two axis interpolation in each of the three planes (XY, XZ and YZ) is carried out, then this type of control is called 2 ½-D path control.

2.4 Types of interpolation (Interpolationsarten)

2.4.1 Linear interpolation (Linearinterpolation)

With linear interpolation, the cutter spindle axis is moved along a *straight line* from one point to the other. If it is required to move along a curved profile as shown in Fig 8.6, the profile can be *broken-up* into a large number of straight lines. The larger the number

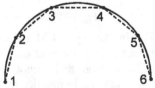

Fig 8.6 Approximation to a curve

of straight lines, the closer the approximation to the curve. When a very close approximation is required, a very *large number of points* need to be calculated and the volume data of data processing becomes enormous. However, the amount of data processing can be reduced considerably by using *other types of interpolation* like circular, parabolic or spline interpolation.

2.4.2 Circular interpolation (Zirkularinterpolation)
Circular interpolation is *limited* in its *usefulness*. It cannot be used to interpolate simultaneous movement along three or more axes. It is also not possible to include rotary axes in circular interpolation. It is principally used for paths which are in the plane of the *principal machine surface*.

3 Drives for CNC machines (Antriebe der CNC Maschinen)
Two types of drives are required for CNC machines, *feed drives* and *main spindle drives*. Both DC and AC *servomotors* have been used for these drives.

3.1 Feed drives (Vorschubantriebe)
Separate feed drives are used to control the movement along each axis of a CNC machine. The feed movements along the axes have to be *extremely precise*, with as little *deceleration* and *overshoot* as possible. All movements should be independent of *opposing forces* such as those due to cutting load, friction or inertia. *Positioning speeds* should be as fast as possible, and movements should be *smooth* and *uniform* without jumps or oscillations.

3.1.1 DC feed drives (Gleichstrom Vorschubantriebe)
DC motor drives are well established for feed drives, and motors fitted with *permanent magnets* to provide the *exciting field* are usually favoured. *Continuously variable speeds* are achieved by *varying* the *armature voltage*. The speed is directly proportional to the armature voltage when the magnetic field remains constant. A current regulator *compensates* for a drop in speed due to *static* or *dynamic* loads by increasing the armature voltage.

3.1.2 AC feed drives (Wechselstrom Vorschubantriebe)
This type of drive uses *a three phase synchronous motor*. The *stator* which produces the exciting field has a normal three phase winding while the *rotor* is fitted with permanent magnets. When a three phase voltage is applied to the stator winding a *rotating field* is created, whose *frequency of rotation* is identical to the frequency of the applied voltage. The rotor which is a permanent magnet rotates at the *same frequency*. The only way the rotor frequency can be changed is by *changing the frequency* of the voltage applied to the stator. This can be achieved by using an *electronic device* called a *frequency converter* which generates a three phase voltage whose frequency can be *varied continuously*. Such motors have several advantages over DC motors like the absence of commutators and carbon brushes.

3.2 Main spindle drives (Hauptspindelantriebe)
3.2.1 DC main spindle drives (Gleichstrom Hauptspindelantriebe)
DC main spindle drives have a *separate excitation winding* in contrast to DC feed drives which have fields generated by permanent magnets. Two operating ranges of speeds are possible, one controlled by the armature voltage which gives *constant torque,* and the other by changing the voltage applied to the field winding which gives *constant power*. Armature control allows a speed ratio of 20 to 30, while field control allows a speed ratio of only 5 to 4.

3.2.2 Three phase asynchronous (squirrel cage) motor drives
(Drehstrom Asynchron-Kurzschlußläufermotoren)

Three phase asynchronous motors with *short-circuited rotors* (squirrel cage motors) have been used for many main spindle drives. In these motors a three phase voltage is applied to the stator winding and this creates a *rotating field*. The rotating field *induces a voltage* in the rotor, causing a *current to flow*, which in turn *causes* the *rotor to rotate*. The *frequency of rotation* of the rotor is however *less than* the frequency of rotation of the rotating field. In other words there is a *continuous slip* which *can vary* with the *load*. This disadvantage can be overcome by using *frequency converters.* Speed ratios of 1:100 can be achieved by using this method.

4 Tool and work changing systems for CNC machines
(Werkzeug und Werkstück Wechselsysteme)

Turret   Disc   Chain

Fig 8.7 Different types of tool magazines

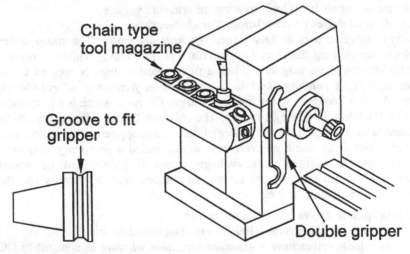

Chain type tool magazine

Groove to fit gripper

Double gripper

Fig 8.8 Tool holder (or adapter)   Fig 8.9 Double gripper used for tool changing

4.1 Tool changing systems (Werkzeug Wechselsysteme)
In CNC machines the manual tool change system has been replaced by *automatic tool changers* and *tool magazines*. The first automatic tool change systems were the *tool turrets* used in capstan and turret lathes, in which each turret had six or eight tools. CNC machines called *turning centres* still have these turrets because they are *cheaper* than other systems. In machining centres, the number of cutting tools required is *much greater* than with turning centres, and disc, chain or rotary type magazines are used (Fig 8.7).

4.2 Tool holders or adapters (Werkzeughalter)
1. The cutting tool is fitted into a *standard tool holder* or adapter (Fig 8.8) which can be *positively locked* into the machine spindle. The tool while it is in its holder is *preset* to *prescribed dimensions* using a set-up which can be far away from the machine.
2. A *positive identification* of the tool can be accomplished by having tool *encoding rings* on the tool holder. More recently identifying has been done by having a *microchip* enclosed within the tool holder body. By using these systems, the tools can be used in *any required sequence* by having the tool number in the program.
3. Tool holders have *grooves* or *gripper slots* (Fig 8.8) in them. These allow automatic tool changers to *change tools* (together with their holders) automatically.

4.3 Automatic tool changing systems
(Automatische Werkzeugwechselsysteme)
An automatic tool change cycle is used to *transfer tools* from the magazine to the spindle of the machine and back. *Double grippers* (Fig 8.9) which are able to grip the tool in the spindle as well as the tool in the magazine *simultaneously* are used. A typical cycle is as follows:
1. Location of the next tool required in the machine.
2. Removal of the last tool to be used from the spindle.
3. Insertion of a new tool into the machine spindle
4. Return of the used tool to the magazine.

4.4 Tool management (Werkzeugverwaltung)
Any program written for CNC machines is based on tools of given dimensions. When the tools used in the machines do not conform to these dimensions, the parts turned out on these machines *will not have the right dimensions*. It is therefore essential that tools are kept in good condition, and *corrections* be made to *compensate* for any *deviations* in tool dimensions.
1. Accurate *presetting* of tools (as mentioned) is a necessary first step.
2. Tool *length compensation* is necessary to allow for the difference in actual length and desired length.
3. Tool *diameter correction* is necessary for cutting path corrections.
4. Tool *wear compensation* is used to compensate for tool wear without changing the programmed offset data.

5. Tool *life monitoring* is used to monitor the usage of each tool and compare this with the permitted cutting life.

4.5 Automatic work(piece) changing systems
(Automatische Werkstückwechselsysteme)

Automatic work changers enable the saving of the *down time* of the *machine tools* caused by the loading and unloading of workpieces, clamping, aligning, etc. This can be done *far away* from the machine without *interrupting* the machining process.

1. Work storing platforms called *pallets* are used. These have *special seating faces* which allow *accurate location* and *clamping* to the work tables of the machine.
2. Pallets are *transferred automatically* from *buffer stations* to the machine tool and back again, after the machining has been carried out.
3. Automatic work changers are necessary if *several work stations* are to be integrated into a manufacturing cell or system.

5 Adaptive control for CNC maschines
(Maximierung der Leistung der CNC Maschinen)

In adaptive control, the objective is to achieve the maximum possible *metal removal rate* within the capability of the machine. CNC machines fitted with adaptive control are able to work closer to their *design limits*, without *overloading* the machine or increasing the complexity of the program.

Adaptive control is a control system which *monitors changes* in *cutting conditions*, and *automatically adjusts* the *spindle speed* and *feed rate* of a CNC machine to produce a component at the *lowest possible cost*.

The programmer when writing a program sets the spindle speeds, feed rates, etc. so that *satisfactory machining* will take place, even under the *worst possible conditions*. This takes care of such *unpredictable variations* as tool wear, variation in the quality of the materials, and variations in the dimensions of the rough workpieces. The program is designed to ensure reliability, repeatability and safety, but *not maximum metal removal*. Adaptive control however automatically changes machining conditions to enable the machine to operate at the *limit of its capability* at all times.

Adaptive control operates by *measuring* such *quantities* as torque, cutting power and motor temperature, by *installing sensors* at suitable places on the machine. The measurements are carried out *continuously* without *interrupting* the machining process. The following changes are identified by adaptive control.

1. Changes in the hardness of the workpiece.
2. Wear of the cutting edges of the tool.
3. Changes in cutting depth.
4. Presence of air gaps in the path of motion.

Changes in *machining conditions* are immediately *compensated* for by changes in spindle speed or feed rate.

6 Programming of CNC machines
(Programmieren von CNC Maschinen)

6.1 Part programming (Manuelles Programmieren)

The term *part programming* means the process of *writing a program* which can be used to machine a *piecepart* (or *workpiece*) on a CNC machine. It *does not mean* the writing of only a *part of a program*. A part program is written using a language which the machine tool can *directly understand*. Computers are not normally used. The usual aids used are pocket calculators and data tables.

A CNC part program consists of a *sequence of blocks* which instruct a CNC machine to carry out a definite machining task. Such a program contains the *geometric data* giving the dimensions and other characteristics of the piecepart (or workpiece) to be machined. It also contains the *path information, switching* and *other instructions* necessary to operate the machine.

6.2 Binary systems (Binärdarstellung)

Digital electronic systems are based on the binary system rather than the decimal system because of the inherent *simplicity* and *reliability* of the binary system.

In the binary system each binary digit or bit has only two states 0 or 1. Larger numbers can be represented in a binary system by having a *combination* of *several bits* as shown in the adjacent figure.

2^3	2^2	2^1	2^0
8	4	2	1

Each position to the left corresponds to an increase in the *power of two*. Thus the binary number 1010 corresponds to the decimal $8 + 0 + 2 + 0 = 10$. Large numbers however require a very large number of bits when represented in the binary system. For this reason a combination of the decimal and binary systems called the *binary-coded decimal system* (BCD) has been devised. This uses a *smaller number* of bits than a pure binary system. In the BCD system only the numbers 0 to 9 are coded as binary, and then the decimal position is used in the same way as in the decimal system. An example is given below.

0111	0010	0110	0100
7	2	6	4

The BCD system is ideally suited to be used on the 8–track punched paper tape shown in Fig 8.10, which is used to feed data into CNC machines. Other forms of data carriers like magnetic tapes and diskettes are also sometimes used instead of punched tape.

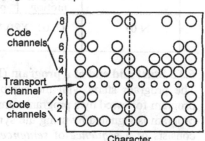

Fig 8.10 8-track punched tape

6.3 Program structure (Programmaufbau)
6.3.1 Words (Wörter)

A word in a program is composed of alphanumeric characters. Each word has *two parts*, *an address* which is an alpha character (a letter), followed by *numeric values*. Each word is an *instruction* to the *control system* of the machine to carry out a certain task. For example, instructions related to the *type* of *movement* are assigned the character G. The word G 01 corresponds to linear interpolation.

Word	⟶	Address	Value
		G	01

Each word can correspond to different kinds of instructions as follows:

1. **Preparatory function path code** – These are instructions relating to the type of motion (G), like rapid traverse (movement) or linear interpolation.
2. **Geometric instructions** – These are instructions are concerned with the relative movement between the tool and workpiece, for example the target positional coordinates X, Y, Z.
3. **Switching instructions** – These are instructions like tool selection (T) and miscellaneous instructions (M) which include items like coolant supply on or off, and spindle on or off. Also included are correction instructions for tool length, zero offsets and cutter radius compensation.
4. **Technical instructions** – These include items like spindle speed (S) and feed rate (F).
5. **Canned cycles** – Instructions that recall parts of a program which are repeatedly used. These are called canned cycles.

6.3.2 Blocks or sentences (Sätze)

A block or a sentence is made up of a number of words. It starts with the letter N and followed by a block or sequence number. The words are arranged in a definite order. An example is shown below.

Block number	Path instructions		Technical instructions			
	Type of motion	Target position	Feed rate	Spindle speed	Tool	Misc: function
N 01	G 01	X20 Y30 Z45	F150	S1200	T02	M07

6.3.3 Start and end of a program (Programmanfang und Programmende)

The program starts with the symbol % which is the start signal for the program followed by a program number, for example % 0035. The end of the program is signaled by using the symbols M 02 or M 30. The program itself consists of a *sequence* of *sentences* which are instructions for tasks to be carried out one after another.

6.3.4 Preparatory function path code G (Wegbedingung G)

The instructions concerning the *motion of the tool* relative to the *workpiece* are specified by the preparatory function path code G, and the path information (or geometric data). The *path code* determines the *type of motion* of the tool, for example along a straight line, or along the circumference of a circle. The instructions consist of the letter G and a two digit number.

The path code is followed by the *geometric data*, for example the coordinate of the target point (X,Y,Z), or the position of the centre of the circle (I,J,K) and the radius R. The distances can be given in absolute or incremental form. When given in absolute form, the distances are related to the workpiece *zero point*. In this case, the path traverse instruction G 90 has to be used. If the incremental form is used, the instruction G 91 has to be used. A selected number of preparatory function path code instructions are given below.

G 00	Fast traverse, point to point positioning	G 42	Cutter radius compensation - offset right
G 01	Linear interpolation	G 53	Cancel zero offset
G 02	Circular interpolation, clockwise	G 54 to 59	Zero shift
G 03	Circular interpolation, Anticlockwise	G 80	Cancel canned cycles
G 04	Dwell (stay)	G 81 to 89	Canned cycles
G 06	Parabolic interpolation	G 90	Absolute input
G 17	XY plane designation	G 91	Incremental input
G 18	ZX plane designation	G 94	Feed rate in mm/min
G 19	YZ plane designation	G 95	Feed rate in mm/ revolution
G 40	Cancel cutter compensation	G 96	Feed rate for const. surface speed
G 41	Cutter radius compensation - offset left	G 97	Spindle speed in rev/min (rpm)

6.3.5 Miscellaneous functions M (Zusatzfunktionen M)

Miscellaneous functions (M) perform a number of *additional tasks* like starting and stopping the spindle or feed, coolant flow on/off,etc. Both preparatory and miscellaneous functions are generally classified as *modal* or *nonmodal*. Modal functions remain effective in succeeding blocks until they are replaced by another function code. Nonmodal functions are only effective in the block in which they are programmed. A short list of these M functions is given below.

M 00	Program stop	M 07	Coolant on
M 02	End of program	M 09	Coolant off
M 03	Spindle on, clockwise	M 13	Spindle (CW) + coolant on
M 04	Spindle on, counterclockwise	M 14	Spindle (CCW) + coolant off
M 06	Tool change	M 60	Piecepart (workpiece) change

6.4 Tool compensation (Werkzeugkorrekturen)

6.4.1 Tool length compensation (Werkzeuglängenkorrektur)

Tool length compensation refers to the correction required when the *actual tool length* is different from the *preset tool length*. The difference between these must be entered into the tool compensation register of the CNC machine.

6.4.2 Tool path compensation

(Werkzeugbahnkorrektur)

The path information in a CNC program usually refers to the required form or *contour* of the *workpiece*. The centre of the cutting tool has to follow a *different path* at a *constant distance* from the contour of the finished workpiece. The path traced by the axis of a milling cutter is shown in Fig 8.11. A *cutter offset* equal to the *radius* of the *milling cutter* is required. The tool path must be *precalculated* by the CNC machine's computer.

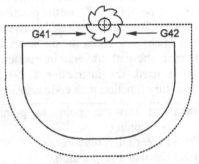

Fig 8.11 Milling cutter compensation

The preparatory function path code instructions G 40, G 41 and G 42 are used for tool path compensation. To choose the right code, the programmer has to look in the feed direction of the tool and observe whether the tool lies to the left or the right of the workpiece contour. The following codes are used depending on the relative position of the workpiece and the direction of motion of the tool.

- G 41 when the tool lies to the left of the workpiece.
- G 42 when the tool lies to the right of the workpiece.
- G 40 cancels the cutter compensation.

6.4.3 Tool nose compensation (Schneidenradius-Korrektur)

The tips of single point cutting tools for lathes, are usually rounded in order to *improve* the *surface finish* of the machined component and *reduce* the *wear* on the cutting tool. In such a case tool *nose compensation* is required if the desired contour is to be generated. A tool whose nose is pointed, and one that has been rounded are shown in Fig 8.12 (a) & (b). When the tool is rounded, it makes contact with the workpiece at *a point B* which is *not* the *theoretical cutting point P*.

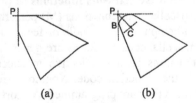

Fig 8.12 (a) Tool with a pointed nose and (b) with a rounded nose

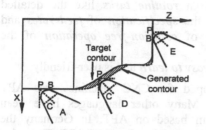
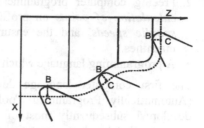

Fig 8.13 Machining errors during   Fig 8.14 Centre point C moves along a path normal to
          a turning process   the tangent contour of the workpiece

- In the case of *feed along the axis* (*axial-feed* shown at A in Fig 8.13) or at *right angles to the axis* (*cross-feed* shown at E in Fig 8.13), no error is made in the cutting process. This is because the contour generating tangent through P is on the same straight line as the theoretical cutting point P, and the straight lines are parallel to the X or Z axes.

- When the feed direction is *not along* the X or Z axis, the generated contour is *different* from the required contour as shown at D in Fig 8.13. The amount of error is dependent on the *angle of inclination* of the tool. The desired shape of contour is machined when the centre point C of the rounded nose of the tool travels along a path which is at a *constant distance* BC *normal* to the *target contour* of the workpiece as shown in Fig 8.14.

A detailed study of part programming procedure is beyond the scope of this book. Those interested would be able to find details in the many comprehensive books on this subject. See for example, Steve Krar and Arthur Gill: CNC technology and programming (McGraw Hill).

6.5 Computer-aided programming (Rechnergestütztes Programmieren)
6.5.1 Initial development (Entwicklungsverlauf)

After the introduction of the first CNC machines, it soon became apparent that for *complex machining tasks,* a *large volume* of *detailed data calculations* had to be carried out before the machining could be done as required. It was necessary to *precalculate* the required data, obtain *additional data* from tables, and to check that there was no *danger of collisions* between tools and workpieces. In order to eliminate these difficulties and to save valuable time and effort, *new programming languages* and *methods* were developed in which the *computer played a major role*. The requirements that had to be met were the following:

1. The possibility of programming *complex machining tasks* in a *simple way*. All detailed data required for carrying out a machining task must be *internally calculated* by the computer from the basic geometric data provided by the programmer.

2. Freeing computer programmers from *routine tasks* like the detailed *calculation of points* on profiles, the *specification* of *feed rates* and *spindle speeds*, and the ensuring of *collision-free operation* of the machines.

3. A programming language which is *easy to learn* and is user-friendly.

The first computer language developed for CNC machines was APT (Automatically Programmed Tools). Many other languages have been developed subsequently most of them based on APT. In Germany the language that has been most frequently used has been EXAPT (Extended Subset of APT). This language is capable of satisfying *both geometrical* and *technological* requirements. In addition to languages like the above, which can be used with any type of machine there are also other *specific languages* which are designed for use with only *one type of machine* or control system.

6.5.2 Processors and postprocessors (Prozessoren und Postprozessoren)

The program which is written in a programming language like EXAPT is not in a form which is suitable to control the machine tool. It is converted into a suitable form in *two successive stages* by *software conversion programs* called the *processor* and the *post processor*. The processor itself is *language-specific* which means that a processor designed for APT cannot be used with EXAPT. The postprocessor however has to be *machine-specific* and has to generate a program which suits a specific machine.

The original program is first converted by the processor into an *intermediate form* called CLDATA (Cutter Location Data*). This form is of a general type* and *is not specific* to any machine tool or controller. The processor, also detects programming and geometrical errors and displays them on the screen. In addition, it performs the *necessary detailed calculations* and calls up any subroutines or *canned cycles* that may be required. Modern processors can also *display* the *progress of machining* on the display screen.

The CLDATA which comes out of the processor is converted by the postprocessor into a *sequence of instructions* specifically designed to *control* the *operation* of a *specific CNC machine tool*. Different types or makes of machines may need *different post processors* and specific machine requirements will have to be strictly followed. Data relevant to the machine itself are usually stored in a suitable form and can be called up when required. Computer aided programming has the following advantages:

1. Use of an *easy symbolic language* to input geometric and technical data.

2. *Reduction* in *geometric data* input with all calculations being done by the computer.

3. Computer *plausibility checks* on the correctness of program and data

4. Graphic *display* of the *geometry* of the part, and also a *graphic simulation* of the *cutting process*.

IX Other manufacturing processes
(Weitere Fertigungsverfahren)

1 Bulk deformation processes (Massivumformprozesse)

Before metal products can be manufactured, the *metal* or *alloy* itself has to be produced. The metal must have the chemical, physical and mechanical properties that are required for manufacturing the desired product. Initially the metal itself has to be produced from *ores* or *scrap metal*. This is done in furnaces by *melting* and *refining*. The *relatively impure* metals or alloys, are normally cast into ingots, billets or slabs. These are metals in their initial form.

Before they can be used for the manufacture of products, these cast metal forms have to be transformed into *secondary* (intermediate or semi-fabricated) *products* like *metal sheets*, *wire* and *rods* from which more *complex products* can be manufactured. Secondary products are produced by *bulk deformation processes* like rolling, drawing, extruding and forging in which large scale *plastic deformation* is involved. Unlike in the sheet *metal pressing* and *forming* processes where the changes in thickness are small, bulk deformation processes cause *large changes* in diameter, thickness, or other dimensions of the metal.

1.1 Strain (or work hardening) caused by deformation (Kalthärten)

When a metal is plastically deformed, it is *strained permanently* and becomes strain or *work hardened*. The metal becomes *brittle*, and *heat treatment* is required to bring the metal back to a softer and more *ductile state*, before it can be used to manufacture metal products.

1.2 Hot deformation processes (Warmumformprozesse)

In hot deformation processes, the metal is deformed after being heated to a temperature that is above the *recrystallisation temperature*. The advantages of the process are:

- The forces and power required to accomplish plastic deformation are smaller because the *metal flows* more *easily* at high temperatures.
- Large deformations are possible without any *danger of fracture*. The generation of *complex shapes* can be done without much difficulty.
- There is *no work hardening* and the components produced are *very strong* and *nonporous*.

Disadvantages are the *oxidation* that takes place with the *formation of scale* on the surface, and the *low dimensional tolerances* of the finished products. Additionally, considerable amounts of energy are needed to heat the product before it is deformed.

1.3 Cold deformation processes (Kaltumformprozesse)

Cold deformation is carried out at room temperature. This process has the following advantages:

- The ability to generate products with *better surface finish*, closer *tolerances*, and *thinner walls*.

- The metal object can be allowed to *retain its strength* in the strain hardened state, or if preferred *heat treated* to bring it to a ductile state.

Disadvantages are that the flow stresses are high, requiring high tool pressure and large amounts of power.

1.4 Rolling (Rollen)

Rolling is the most important of the bulk deformation processes. In flat rolling, the thickness of a cast slab is reduced resulting in a product that is longer and thinner, but only slightly wider. Cast slabs are *initially rolled* by *hot rolling* processes. The sheets produced have *poor dimensional tolerances* and rough surfaces. These sheets are relatively thick, and are used in such applications as boiler making, ship building, and in the construction of welded machine frames.

Thinner sheets are produced by cold rolling the thick sheets produced by hot rolling. Cold rolled sheets have *closer tolerances* and a *better surface finish*. *Shape rolling* is used to manufacture rods, long bars, etc. each having a uniform but different cross-section.

In processes like ring rolling (Fig 9.1(b)), and tube rolling (Fig 9.1(f)) which are used to manufacture *hollow products*, pierced billets and centred mandrels are used. Screws, taps, etc. can be produced by *thread rolling* processes. These products have screw threads which are *stronger* than those produced by thread cutting.

1.5 Drawing (Ziehen)

In this process, the metal is *pulled through a die* whose cross-section gradually decreases. Most wire types of circular, square or other cross-sections are manufactured by drawing processes.

Many products like *nails, screws, bolts* and *wire frames* are made from metal wires. *Seamless tubes* are also manufactured by drawing processes. Large diameter tubes are usually manufactured by hot drawing, but small diameter tubes below a certain diameter must be cold drawn.

1.6 Extrusion (Strangpressen)

The extrusion process can be used to produce long tubes and rods of *uniform cross-section*. In this process, the metal is under pressure and is *forced to flow* through a die. The cross-sections of the extruded products can have different shapes and sizes, depending on the shape and size of the opening in the die.

- In direct or forward extrusion, the extruded metal and the punch which pushes the metal move in the *same direction* (Fig 9.1(g)).
- In indirect reverse (or back) extrusion, the extruded metal and the punch move in *opposite directions* (Fig 9.1(h)).
- Hot and cold extrusion processes are both possible, and the extrusion of *hollow products* is carried out by using a *centred mandrel* (Fig 9.1(i)).

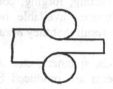

Fig 9.1(a) Flat rolling

Fig 9.1(b) Ring rolling

Fig 9.1(c) Form rolling

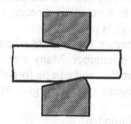

Fig 9.1(d) Wire drawing

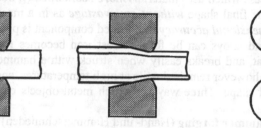

Fig 9.1(e) Tube drawing

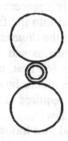

Fig 9.1(f) Tube rolling

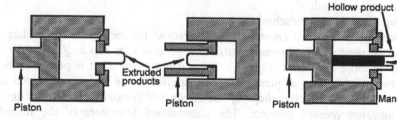

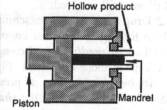

Fig 9.1(g) Forward extrusion   Fig 9.1(h) Reverse extrusion   Fig 9.1(i) Extrusion of hollow products

2 Forging (Schmieden)

Metal components can be produced in many ways, by casting, forging, cold forming, welding, etc. Casting is probably the cheapest process available, but cast components are often *porous* and also *brittle*, with the result that they *break easily*.

Some metal components like crankshafts, connecting rods, spanners, etc. are subjected to *severe stress* when in use. Such components are produced by forging, because forged components are *stronger* than those produced by casting or machining from solid bar material. This is because the *grain structure* in these cases is different. Each *grain* in the metal is a *crystallite* (or a tiny crystal) and the strength of the metal depends on the *size, shape* and *orientation of the grains*. The *difference in grain structure* between a gear tooth that has been machined and one that has been forged is shown in Fig 9.2.

Forging is a process which uses materials *more economically*, because the metal is pressed into the final shape *without any wastage* as in a machining process. However the *dimensional accuracy* of a forged component is poor.

Not all metals and alloys can be forged. Cast iron becomes *very brittle* when heated to red heat, and breaks easily when struck with a hammer. Many steels and other metals however remain ductile at high temperatures, and can be forged into the required shape. Three ways in which metal objects can be forged are given below.

2.1 Hand and hammer forging (Hand- und Hammerschmieden)

Over the centuries, blacksmiths produced small quantities of forged components using hand tools. The red hot metal was forged into the required shape by the blacksmith who was *a highly skilled person*. No dies were used in this process. The forging of much larger components was possible by the use of large *mechanical hammers* after the beginning of the industrial revolution. The power required to operate large hammers became available, but a large degree of skill was still required.

2.2 Drop forging (Gesenkschmieden)

Drop forging is the most *commonly used method* for the production of large numbers of *medium sized components*. This process uses *two half dies* which are kept in close alignment (Fig 9.3). The red hot metal billet is placed in the cavity between the dies. If pressure is now applied to the top die, the metal is forced to flow and take the *shape of the cavity* which exists between the dies, when they are pressed together. The dimensional tolerances of the forged component are usually poor, and the component usually needs to be subjected to a succession of trimming and machining processes before it can be used.

2.3 Upset forging (Schmieden mit Axialdruck)

In this process, a *part of the metal* used to form the component is heated to *red heat*, and *axial pressure* is applied to *change the cross-section* of the component. Several stages are required to produce a component (Fig 9.4).

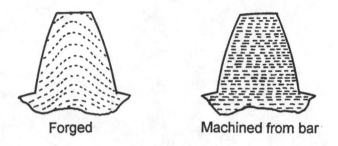

Forged   Machined from bar

Fig 9.2 Difference in grain structure between a forged and a machined gear tooth

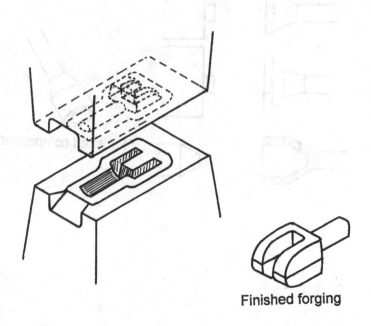

Finished forging

Fig 9.3 Drop forging produced by using split dies

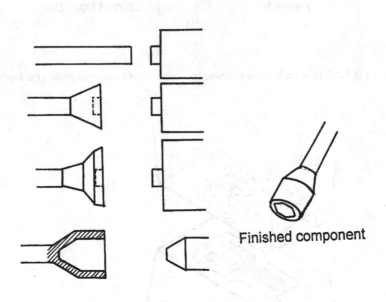

Finished component

Fig 9.4 Several stages in the production of a socket spanner by the method of upset forging

3 The casting of metals (Gießen)

Metal objects have been produced by the process of casting for thousands of years. This is probably the *cheapest* and *quickest* way of producing a metal object, particularly a large and *complex one*. There are many ways of producing a casting, some of which are described below.

3.1 Sand casting (Sandguß)

Sand casting was probably the *oldest way* of producing a metal casting. In this method, the metal is melted and poured into a *hollow space* in a box filled with sand. The box with its hollow space is called a *mould*, and the hollow space has approximately the *same shape* and *size* as the object which has to be produced as a casting. The production of a casting is usually done in four stages as follows:

1. Making of the pattern
2. Making of the mould
3. Pouring in of the metal from the mould
4. Removal of the casting from the mould

The *pattern* is a replica of the object to be cast and is usually made from metal or wood. It is slightly larger than the object to be cast. A wooden pattern is made of *two halves* which can be joined together by using *dowel pins* (Fig 9.7).

Some castings are *completely solid*, while other castings are *partly hollow*. *Hollow castings* are known as *cored castings*. The hollow part or core is separately made as shown in Fig 9.6 and placed in the mould. A mould and a core which are ready for the pouring in of the metal are shown in Fig 9.8.

The molten metal which is poured into the mould *shrinks slightly* on solidifying. The pattern is usually made *slightly larger* than the final form of the object to allow for *shrinkage*, and also so that the casting can be *machined* to *definite dimensions*. These extra allowances in the size of the casting are called *shrinkage* and *machining allowances*. Large and complex objects like engine blocks, or beds of machine tools are usually made by the sand casting process.

3.1.1 Sands for moulds (Modellsand)

The *cheapest sand* used for moulds is silica sand (SiO_2) and if its composition and contamination level are carefully controlled, it is usable at the *highest casting temperatures* including that for steel. Other special sands used are zircon, chromite or olivine. The sands need to be *bonded* by using a bonding agent or *binder* to withstand the pressure of the molten metal, and also the *erosion* caused by the metal. The bonded sand must be sufficiently *porous* to allow the *gases* present in the molten metal *to escape*. Sands are tested for tensile and compressive strength, shear properties, porosity and compactability.

Green sand moulds are bonded with clay, but sometimes a binder like dextrine is used which hardens the surface.

Dry sand cores are made from silica sand and a binder which is usually an oil which hardens when heated.

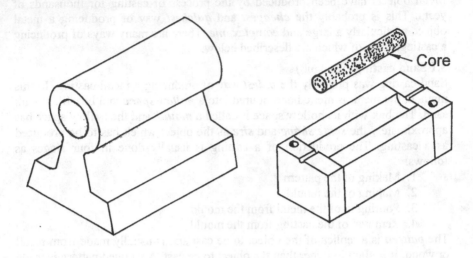

Fig 9.5 Finished casting with a cored hole

Fig 9.6 Method of producing cores

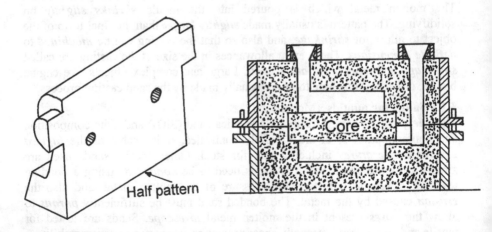

Fig 9.7 One half of a wooden pattern

Fig 9.8 A mould ready for pouring in metal

Carbon dioxide moulds use *silica gel* as a bonding agent. When the mould is finished, carbon dioxide gas is passed through the sand. This produces a hard mould which does not need baking.

Oil sands consist of sand mixed with a *vegetable oil* such as linseed oil, and some *cereal flour*. On heating the oils form a *polymer,* which ensures high strength bonding. These sands are particularly suitable for cores.

Shell moulding is a process which uses a *synthetic resin binder* and produces castings with a *smooth finish* and *close dimensional tolerances.* The mixture of sand and resin is placed over the metal pattern which is heated to a temperature between 200°C and 260°C. The resin in the mixture forms a *thin shell* over the pattern. When the shell has reached the required thickness, the sand is removed by rotating the pattern which results in the sand being thrown out. The remaining shell is cured on the pattern and then removed. *Mating halves* of the shell are *combined* with a cement and surrounded with sufficient *backing material.* The casting is now made by pouring the metal into the resin mould formed.

3.2 Permanent mould casting (Gießen in Dauerformen)

In sand casting, the mould is destroyed after the casting has been made. In the permanent mould casting processes, the *mould can be used repeatedly.* The mould is made from a material which has the following properties:

1. A sufficiently *high melting point* to withstand erosion by the metal.
2. A *strength* that is high enough to prevent deformation under repeated use.
3. A high *thermal fatigue resistance* that prevents the formation of fatigue cracks.
4. A low *level of adhesion* to the molten metal. This prevents the metal from welding itself to the mould.

Mould materials used are cast iron, alloy steels, molybdenum alloys and also graphite.

Coatings composed of refractory powder in a suspension are applied to the die surface as *protection* for the surface and to reduce heat transfer.

Parting compounds consisting of graphite, silicone, etc. are used to *reduce adhesion* and help *ejection* of the casting from the mould.

Metals suitable for casting by this method are alloys of zinc, aluminium, magnesium and lead, certain bronzes and cast iron. The castings cool rapidly in the permanent moulds, and therefore have a dense *fine grained structure* which is free from *blow holes* or *shrink defects.* Better surfaces and closer tolerances are obtained in comparison with sand castings.

3.2.1 Gravity die casting (Gießen mit Schwerkraft)

This process uses *permanent moulds* which are made as *two halves* which can be assembled together during casting and moved apart for removal of the casting. The molten metal is poured into the mould and fills it under the action of gravity. Automatic machines have *hydraulic actuators* and *automatic feed systems* to feed in the metal.

3.2.2 Pressure die casting (Druckgießen)

In this process, the molten metal is forced into the mould cavity by *external pressure*. The metal is forced hard against the surface of the cavity producing a very *accurately made casting* with an *excellent surface finish*. Such castings are superior to those made by the gravity die casting process and do not need to be subjected to *further machining* or *finishing* processes. There are two types of pressure die casting processes (a) the hot chamber process and (b) the cold chamber process

(a)The hot chamber process (Warmkammerverfahren)

The goose neck machine (Fig 9.8 (b)) used in this process, has a goose neck which dips into the melting pot. The liquid metal is transferred to the mould directly from the melting pot by a *pump* consisting of a *cylinder* and a *plunger*.

(b) The cold chamber process (Kaltkammerverfahren)

Here the metal is melted outside the machine and sufficient metal for each casting is poured in manually or automatically into the shot chamber. The plunger moves and pushes the molten metal into the mould. In all the above die casting processes, the casting is *ejected from the die* after solidifying and the machine *automatically proceeds* to make the *next casting*.

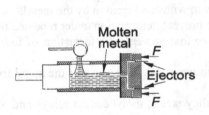
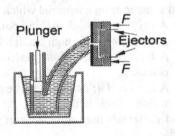

Fig 9.9 (a) Cold chamber pressure die casting   Fig 9.9 (b) Hot chamber pressure die casting

3.2.3 Centrifugal casting (Schleudergießen)

In centrifugal casting, a portion of molten metal is poured into a hollow water cooled metal tube (lined with refractory material) which acts as a mould. The mould is *rotated at high speed*, and the *centrifugal forces* cause the molten metal to be *pressed tightly* against the mould. The metal solidifies quickly due to the water cooling, and a sound dense casting in the form of a tube is produced.

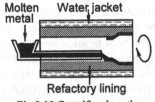

Fig 9.10 Centrifugal casting

4 Shearing and blanking (Scheren und Stanzen)

The terms cutting, shearing and blanking refer to some of the ways of separating a sheet of material into parts.

4.1 The cutting process (Messerschneiden)

In a cutting process a knife whose cutting edge is shaped like a wedge is *pressed* against a material to *separate it into parts* (Fig 11(a), (b) & (c)). This method of separation can only used with *soft materials* like paper or leather.

4.2 Shearing (Scherschneiden)

This is the term applied to a process where the separation of the material takes place between two cutting edges which pass each other (Fig 9.11(d)). *Hand shears* are used for cutting pieces of material by hand. The cutting blades are *hollow ground* so that when cutting takes place, the blades touch each other only at one point. *Guillotines* are used for the cutting of large sheets, while *presses* can be used for the blanking of complex forms.

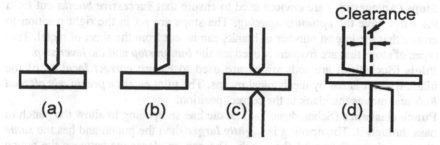

Fig 9.11 (a), (b) & (c) Cutting process   Fig 9.11 (d) Shearing process

4.2.1 Types of shearing operations (Verfahren beim Scherschneiden)

1. **Blanking (Stanzen)** refers to the *removal* of a *piece of material* of the desired shape from a larger sheet. The removed piece is in this case more important than the hole produced.
2. **Piercing (Lochen)** refers to the production of a hole of *any shape* in a sheet of metal by the use of a punch and a die. The *removed piece* of material is *not important* and is regarded as scrap.
3. **Punching** is a special kind of piercing where a *circular hole* is produced.
4. **Notching (Ausklinken)** refers to the removal of a piece of metal from the *edge* of a metal sheet.
5. **Lancing (Einschneiden)** refers to the process of *partially cutting* through a sheet of metal without removing any material.
6. **Slitting (Abschneiden)** refers to the operation of cutting a metal sheet along a *straight line* parallel to its length.
7. **Perforating (Perforieren)** refers to the production of a *regularly spaced array of holes* in a sheet of material.

8. **Nibbling** (Knabberschneiden, Nibbeln) refers to the use of *repeated small cuts* in order to remove a piece of metal from a metal sheet.
9. **Trimming** (Beschneiden, Trimmen) is a term used to describe the removal of *excess material* from a pressing (or pressed object).

4.3 Press tools (Schneidwerkzeuge)

Press tools (or die sets) consisting of *punches* and *dies* are used for the blanking and piercing of *large numbers* of *metal parts*. The basic construction of one type of press tool is shown in Fig 9.12.

Sliding pillars (Führungssäule) are used to *guide the punch* precisely as it moves downward into the die.

Stripper plate (Abstreifer) – The stripper plate is *mounted above* both the die and the metal strip from which the blanks are produced. The stripper plate has an opening large enough for the punch to *pass freely through*. As the punches rise after the downward stroke, the metal may be lifted. The stripper plate *prevents* the *strip* from *rising too far*.

Stops (Anlagestifte) are devices used to ensure that *successive blanks* cut from the sheet have the *optimum spacing*. The stops are set in the right position to ensure that the largest number of blanks can be cut from the sheet of metal. Two types of stops that are frequently used are the *button stop* and the *lever stop*.

Pilots (Suchstifte) are rods which are used to ensure *correct location* of the blank when it is fed by mechanical means. The pilot enters a *previously pierced hole* and moves the blank to the correct position.

Punch clearance (Schneidspalt) – The die has an opening to allow the punch to pass through it. The opening is *slightly larger* than the punch and has the *same shape* as the periphery of the punch. The gap or *clearance* between the punch and the top or cutting edge of the die has to be carefully controlled. Its value depends on the thickness and the shear modulus of the sheet which is being blanked. The clearance varies between 0.5 % and 5 % of the sheet thickness.

4.4 Blanking (Stanzen)

Blanking can be done by using a press that is fitted with an appropriate die set. The *double blanking die set* shown in Fig 9.12 consists of two punches and a die. The punches which have *sharp edges* at the bottom are attached to a punch holder which is fitted with two guide collars. The collars can move vertically on two pillars attached to the die set. The punches are *slightly smaller* than the holes in the die.

The metal in the *form of strip* is fed in from the right side, up to the *small stop stud*. The punches now move down *through* the *stripper plate*, and two blanks are punched from the strip. As the punches rise after the downward stroke, the metal strip will be lifted, but is *prevented from rising* too far by the stripper plate. The metal strip is next *pushed forward* by an *appropriate distance* determined by the *stop stud*, and the punches move down to produce two more blanks. The process can be continued as long as necessary.

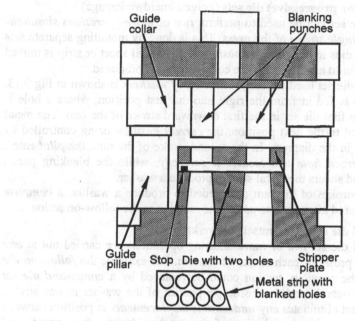

Fig 9.12 Blanking die set with two punches

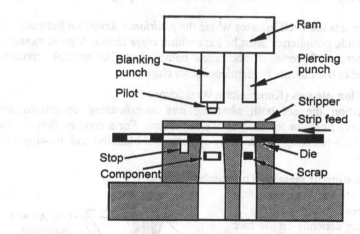

Fig 9.13 Combined blanking and piercing using a follow-on die set

4.5 Follow-on (or progressive) die sets (Folgeschneidwerkzeuge)

A follow-on die set can be used to perform *two* or *more operations simultaneously* with *a single stroke* of the *press*. This is done by mounting separate sets of punches and dies in *two* or *more positions*. The metal sheet or strip is moved from one position to another, until the complete part is produced.

A two position die set used for the *production of washers* is shown in Fig 9.13. The metal strip is fed in from the right into the first position, where a hole is produced by the first die set in the first downward stroke of the ram. The metal strip is advanced to the next position, the *correct position* being controlled by the *stop* shown in the diagram. In the second stroke of the ram, the *pilot* enters the already *pierced hole* and *locates it correctly*, while the blanking punch moves down and shears the metal strip to produce a washer.

Although two strokes of the ram are needed to produce a washer, a *complete washer* is delivered after *each stroke* of the ram due to the follow-on action.

4.6 Compound die sets (Gesamtschneidwerkzeuge)

In a compound die set *two or more* shearing operations are carried out at *one position* of the press for each stroke of the ram. For example the *follow-on die set* shown in the previous section could be replaced by a *compound die set* which cuts the outside and the inside peripheries of the washer in one stroke. This arrangement eliminates any *undesirable displacements* in position between the inside and outside peripheries of the washer due to *inaccurate feed movements* between the two strokes or due to *sideways play* of the metal strip in the feed track.

Compound die sets are used in cases where the *positional deviation* between the inside and outside peripheries must be kept within *close limits*. A good example is the manufacture of precision mechanical components like multiple contacts which are used in the mounts for semiconductor chips.

4.7 Combination die sets (Kombinierte Werkzeuge)

In a combination die set both *shearing and nonshearing* operations are performed at *one position* for each stroke of the ram. For a combination die set can carry out both blanking and drawing operations if blanking and drawing dies are built into it.

5. Thermal cutting of metals (Thermisches Trennen)

5.1 Oxygen cutting

(Autogenes Brennschneiden)

Oxygen cutting depends on the fact that *iron oxidizes rapidly* in the presence of oxygen at a temperature of 816°C. This temperature is well below the melting point of iron.

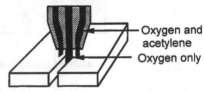

Oxygen and acetylene

Oxygen only

Fig 9.14 Oxygen cutting

This process can only be used only with iron and steel products which have a low alloy content. It can be used to cut steel products such as sheets, bars, pipes, forgings, castings, etc. The type of burner used is shown in Fig 9.14. A mixture of oxygen and acetylene flows through the outer tube and when lighted prewarms the metal at the cutting point. The stream of oxygen coming out of the central tube hits the central spot. The steel is oxidized and the resulting slag is carried away by the stream of oxygen and a narrow clean cut is produced.

5.2 Plasma cutting (Plasmaschneiden)

Plasma cutting is a process that can be used to cut alloy steels and nonferrous metals. The oxides formed by these materials have melting temperatures which are higher than the melting temperatures of the metals themselves. The type of burner used is shown in Fig 9.15. This process uses a tungsten electrode.

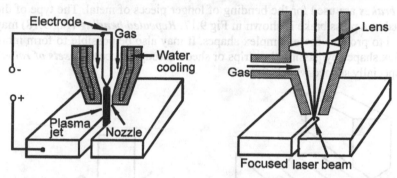

Fig 9.15 Plasma cutting   Fig 9.16 Laser cutting

A pilot arc is first produced between the electrode and the nozzle by using a high frequency source. If the gas is now allowed to flow past this arc, the gas becomes ionized and an arc is now struck between the electrode and the workpiece. This arc soon becomes a high density plasma jet with temperatures of above 20,000 °C. The thin jet melts a thin portion of the metal and flushes it away. Ordinary carbon steels are cut by using air or oxygen, while nonferrous metals and stainless steels are cut by using a mixture of hydrogen and argon, or hydrogen and nitrogen.

5.3 Laser cutting (Laserschneiden)

When the radiation from a laser beam is focused, the intensity can be great enough to cause the melting of a metal (Fig 9.16). A lens is used to focus the light from a laser beam to a spot having a diameter between 0.1 and 0.3 mm. The energy density which reaches a level of about 10^7 W/cm^2 quickly melts the metal. The melted metal is carried away by a stream of gas. A 1 kW laser can cut through 1m thick steel sheets. Laser cutting is a process that can be used to cut almost any type of metallic or nonmetallic material.

6 Bending and forming processes (Umformverfahren)

6.1 Introduction (Einführung)

Many products are manufactured by bending and forming sheet metal into various shapes. These are *cold working operations* in which the sheet metal is *plastically deformed*. The sheet metal has to be stretched beyond the *elastic limit*, but care has to be taken that the stretching does not go so far as to cause *cracking* or *fracture* of the metals. Metals suitable for this kind of work are low carbon steels, killed steels and alloys of copper or aluminium.

6.2 Forming by bending (Biegeumformen)

In some cases, *simple bending* of a metal sheet is all that is necessary to yield the required product. Die sets in mechanical presses are used to bend small pieces of metal into *complicated shapes*. Special presses with long beds called *press brakes* are used for the bending of longer pieces of metal. The type of die set used in a press brake is shown in Fig 9.17. *Repeated bending* (Fig 9.18) may be used to produce more complex shapes. It may also be possible to form more complex shapes by passing the strips or sheets through successive *sets of rollers* with specially shaped wheels.

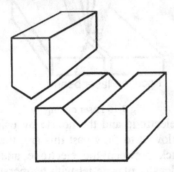

Fig 9.17 Die set for a press brake

Fig 9.18 Repeated bending
using a press brake

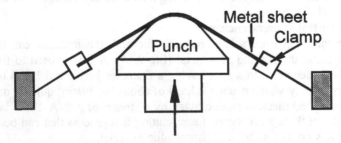

Fig 9.19 Stretch forming using a punch without a die

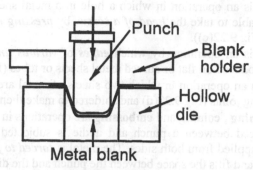

Fig 9.20 Stretch drawing

6.3 Stretch forming (Streckziehen)

In the stretch forming process, the metal sheet is *clamped round* its *periphery*. This process uses *only a punch* to stretch the clamped sheet (Fig 9.19). The change in shape causes the sheet to stretch and become *thinner*. An example of *stretch drawing* is shown in Fig 9.20.

6.4 Deep drawing (Tiefziehen)

In this process, the metal blank which is *not clamped* is allowed to draw into the die. A number of stages are required satisfactory deep drawing. The first stage which is shown in Fig 9.21 is called *cupping*.

Initially the pressure pad presses firmly on the die. When the punch moves downwards, the blank is pushed into the cavity. The metal bends and *flows plastically* when it is drawn into the cavity to form a *cup*. Any *wrinkles* formed at this stage are *ironed out* by the *pressure pad*. *Several stages* of drawing are required to produce deeper objects. Simultaneous blanking and drawing can be carried out by using combination dies.

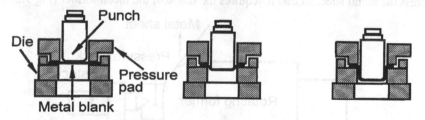

Fig 9.21 First stage in a deep drawing process (also called cupping)

6.5 Other forming operations (Weitere Umformverfahren)

- **Beading** is an operation which improves the appearance, safety, strength and stiffness *by folding over* the edge of the metal sheet (Fig 9.22(a).
- **Curling** is a similar operation to beading in which the edges of a metal sheet are *curled* or *rolled* over (Fig 9.22(b)).

- **Plunging** is an operation in which a hole in a metal sheet is *bent* into a shape suitable to take the *head of a screw,* by *pressing a punch* through the hole (Fig 9.22(c)).
- **Flanging** is an operation by which *edges* of *various angles* and *widths* may be produced on flat or curved metal sheets or tubes (Fig 9.22(e).
- **Folding** is an operation in which two sheets of metal are folded to form interlocking joints (Fig 9.22(d) and soldered to make them water-tight.
- **Cold forming** , **coining** and **embossing** are operations in which a piece of metal placed between a punch and a die is subjected to a *very high pressure* applied from both sides. The metal is *forced to flow* while in the cold state and fills the space between the punch and the die.

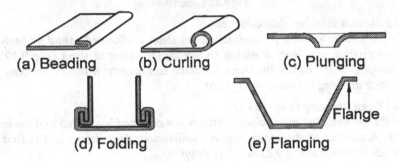

(a) Beading   (b) Curling   (c) Plunging

(d) Folding   (e) Flanging

Fig 9.22 Some bending and forming processes

6.6 Spinning (Drücken)

In this process, a thin sheet of metal is *revolved at high speed and pressed against a former* which is attached to the spindle of a lathe. The metal is supported through a pressure block at the tailstock. A *special tool* is used to press the metal sheet so that it acquires the shape of the metal former (Fig 9.23).

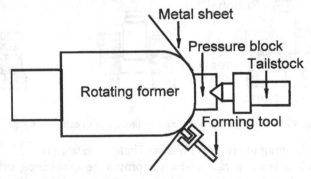

Fig 9.23 Metal spinning

7 The coating of surfaces (Beschichten)

7.1 Introduction (Einführung)

Surface coatings are an important part of many products and a variety of coatings are used for purposes of *decoration, surface protection, improving corrosion resistance,* and for improving the *wearing qualities* of the surface.

It is at first necessary to *clean the surface* thoroughly so that it is free of dirt, oil, grease, scale, oxides, etc. which affect the adhesion and the life of the coatings. The different types of coatings may be divided into the following groups:

1. Metallic coatings
2. Organic coatings
3. Inorganic coatings
4. Chemical conversion coatings

7.2 Metallic coatings (Metallische Überzüge)

Metallic coatings may be applied in many ways, by electroplating, hot immersion, chemical deposition or by the spraying of hot metal (metallizing).

7.2.1 Electroplating (Galvanisieren)

In this process, a current is passed between two electrodes, an anode and a cathode which are placed in a *solution of a salt* (or compound) of the metal which forms the coating. The metal object to be plated is attached to the *cathode*, while the metal which forms the coating is attached to the *anode*.

(a) **Zinc plating** is the cheapest and most commonly used form of plating. It is mainly used to plate steel and iron objects, and is able to provide them with good *protection* against *corrosion*.

(b) **Cadmium plating** is more expensive than zinc but has the advantage of providing better *corrosion protection* under *salty conditions*.

(c) **Nickel plating** provides a surface which has excellent *wear* and *corrosion resistance* and the ability to be joined by soldering or brazing.

(d) **Tin plating** provides a good surface, but is relatively expensive to make. It is mainly used for coating tin cans, kitchenware and food containers because of good *tarnishing resistance*, and also because of the *nontoxic nature* of the oxides formed.

(e) **Copper plating** has decorative value, but has to be protected from oxidation by a clear lacquer.

(f) **Silver plating** is used mainly for decorative purposes. Disadvantages are its high cost and its susceptibility to *tarnishing*.

(g) **Gold plating** is used mainly for decorative purposes and for its excellent resistance to tarnishing. It is used for plating jewellery, watches, etc. and for plating electronic components and circuits.

(h) **Chromium plating** is used because of its decorative properties and also for producing *wear-resistant surfaces* on tools. piston rings, etc.

7.2.2 Hot-dip coating (Schmelztauchüberzüge)

This is a rapid and inexpensive process that is cheaper than electroplating. *All parts* of the object to be coated including joints and crevices *can be covered*. The commonest process is *galvanizing* (or hot-dip galvanizing), in which thoroughly cleaned objects made of iron or steel are dipped in a bath of *molten zinc*. Zinc gives electrochemical protection against rusting, and can be effective in preventing rust even when part of the coating is removed. Aluminium, tin and lead coatings can also be applied by the hot-dip method.

7.2.3 Metallizing (or thermal spraying) (Metall-Spritzüberzüge)

In this process, metallic or sometimes nonmetallic coatings can be deposited by *spraying fine globules* of the coating material on the object to be coated. The metal is initially in the form of rod, wire or powder. The commonly used flame spray process uses a spray gun that feeds the metal wire through a nozzle surrounded by an oxy-acetylene flame. A stream of air breaks the molten metal into globules and sprays it on the surface.

The sprayed metal is *porous* and can *absorb oil*. Bearing surfaces from Babbitt metal can be built by metal spraying. Sprayed coatings are used for building up *worn components*, for *corrosion protection* and for improving the *wearing qualities* of the surfaces. Paper and cloth are coated with sprayed metal for use in electrical capacitors.

7.3 Organic coatings (Organische Überzüge)

Organic coatings may be divided into three classes – paints, varnishes and lacquers. These usually have three main constituents:

- A vehicle binder (or filmogen) like a drying oil or a resin.
- A pigment
- A thinner

In addition *other ingredients* (often called additives) like plasticizers, catalysts, emulsifiers, antifoamers and thickeners are added as required.

Pigments may be natural or synthetic, organic or inorganic, opaque or nonopaque. Inorganic pigments like white zinc oxide or titanium oxide, red iron oxides or red lead (for rust prevention), yellow iron oxide or zinc yellow, blue ultramarine, etc. are used in addition to organic dyes.

7.3.1 Paints (Lacke)

- a) **Oil paints** contain oil as a vehicle, and are mainly used for exterior surfaces. They need a relatively long time for drying.
- b) **Enamel paints** are harder, glossier and smoother than other types of paints. This arises from the use of resin or varnish as the vehicle. Enamels are widely used as organic coatings in the metal processing industry, because of their availability in a large range of colours, their resistance to corrosion, and their ease of application.
- c) **Baked enamels** have a harder finish, which is more abrasion resistant than typical air-drying enamels.

d) **Bituminous paints** are used to protect metal and masonry where there is no objection to their black colour. They contain hard asphalts cooked with drying oils.

7.3.2 Varnishes (Klarlacke)

A varnish is a mixture of resin and drying oil dissolved in a volatile thinner. Varnishes may be clear or may contain a dye, but they do not hide a surface when applied to it. The addition of a *pigment* produces an *enamel* which hides the surface.

a) **A spirit varnish** is a solution of resin alone in a thinner, for example shellac varnish.

b) **An oleoresinous varnish** consists of a mixture of resin and drying oil dissolved in a thinner.

c) **A spar varnish** is made of a mixture of phenolic resin, dehydrated castor oil and linseed oil dissolved in a thinner.

7.3.3 Lacquers (Schnell trocknende Lacke)

Present day lacquer refers to quick drying coatings which contain cellulose acetate, cellulose nitrate, or cellulose acetate butytrate. In addition they contain resins, plasticizers and solvent.

7.3.4 Special paints (Speziallacke)

Numerous types of paints are produced which are required for a special purpose, have a special finish, and have special ingredients and formulas. Among these are paints resistant to chemicals, emulsion paints, powder coatings, fire-retardant paint, marine paint, fungicides, wood preservatives, crackle finish paint, hammer finish paint, etc.

7.4 Inorganic coatings (Anorganische Überzüge)

Inorganic coatings mainly contain refractory materials. They have an attractive finish and also good resistance to corrosion and oxidation. They have surfaces which are hard, rigid, abrasion resistant, and thermally insulating. They can also resist high temperatures.

a) **Porcelain enamels** have surfaces with the *strength* and *stability* of *steel* combined with the *beauty* and *usefulness* of *glass*. They are durable, easy to clean, and have good colour stability.

Porcelain enamels can be applied in many ways, by dipping, by manual and electrostatic spraying, flow coating and by the application of dry powder. The coated objects are fired at temperatures of about 600°C. Conveyer equipment can be used for successive operations like spraying, dipping, drying and firing. Very often a *single coat* is sufficient, and this is usually of good quality and inexpensive to make.

b) **Ceramic coatings** are particularly useful as *protection* for *surfaces* that are subjected to elevated temperatures. In addition to protecting metal surfaces from oxidation and corrosion, ceramic coatings increase their strength and rigidity. The materials used in the coatings are mainly silicate powders, but carbides, silicides and phosphates may also be used. The coatings may be applied by spraying, dipping, flow coating, etc.

7.5 Chemical conversion coatings
(Beschichten durch chemisches Abscheiden)

Chemical conversion coatings are produced when a film is deposited on the surface as a result of a chemical reaction. The most important types of coatings are described below.

a) **Phosphate coatings** are mainly used as a *base* for the *application* of *paint* or enamel. The surface is treated with a dilute solution of phosphoric acid and other ingredients to form a *mild protective layer* of crystalline phosphate. This process is widely used in the automobile and electrical appliance industries.

b) **Chromate dip coatings** are used as an added *corrosion protection* for zinc or cadmium coated steel sheets. They are also used for protecting objects made of nonferrous metals like aluminium and magnesium. The coatings are very thin and also as a base for painting. The object to be coated is dipped in a solution of chromic acid mixed with other acids and salts. The chemical reaction produces a protective film containing chemical compounds. Coloured coatings can be produced by adding suitable organic dyes.

c) **Anodic coating** refers to the process of forming an *oxide coating* on the surfaces of metals like aluminium and magnesium. Here the object whose surface is to be anodized is made the *anode* in an *electrolytic bath* containing chromic and other acids. An oxide coating is formed which protects the metal from corrosion and acts as a base for painting. This process is widely used in the aircraft industry. Excellent *coloured coatings* may be produced by immersing the coated objects in warm dye solutions and then sealing the dye in the porous coatings by dipping in dilute nickel acetate.

8 The manufacture of plastic goods
(Herstellung der Kunststoff Produkte)

8.1 Introduction (Einführung)

Plastic products are of recent origin, but they are being increasingly used, particularly as replacements for metal products. This is because they are cheap, light, easy to manufacture, and easy to maintain.

Processes used in the manufacture of plastic goods are very *cost-effective*, because products having complicated shapes can be produced in a *single operation*, with no further work having to be done on them. However the dies used in these processes have to be very *precisely made* and must have an *excellent finish*. This makes the dies very expensive to make, and the processes can only be profitable if *large quantities of goods* are produced. The raw materials used in manufacturing these goods are usually in the form of powder granules or liquid. The plastic materials which are most commonly used are of two types, thermoplastics which can be moulded several times, and thermosetting plastics which can be moulded only once.

8.2 The injection moulding process (Spritzgießen)

This is a process that is used for the manufacture of thermoplastic goods and is probably the *most widely used* process in the manufacture of plastic products. The raw material which is in the form of granules, is fed into a heated plasticizing chamber through a screw feed mechanism (Fig 9.24).The raw material is first heated, compressed and degassed, until it is in a soft state. The screw feed mechanism is next given a sudden push forward. This forces the soft plastic through an injection nozzle into a two piece mould.The mould is next cooled rapidly, causing the plastic in the mould to harden quickly. When the plastic has hardened, the mould opens, and the finished component is ejected from the mould.

8.3 The extrusion process (Extrudieren)

The extrusion process is commonly used for the production of bar, tube, sheet, etc. from thermoplastic materials. Extrusion machines have screw feed mechanisms similar to those fitted be to injection moulding mchines. The soft plastic is forced through a die (which has the desired cross-section) in the form of a strand, and is hardened by cooling in a stream of air. Bars, sheets, tubes, etc. are *semifabricated products* which are used to fabricate more complicated products.

8.4 Thermoforming processes (Warmumformen)

Open container like objects can be made by thermoforming processes from thermoplastic sheets which are heated to temperatures of between 60° and 90°C.

   a) In the vacuum forming process, the sheet is *clamped* round its *periphery* and heated. When the sheet has become soft, a vacuum is applied from below and the *sheet is drawn* into the *female die*. On cooling the plastic sheet has the form of the die (Fig 9.25).

   b) In pressure assisted vacuum forming, air pressure is applied from above in addition to the vacuum applied from below (Fig 9.26).

8.5 Extrusion blow moulding (Druckumformen)

In this process, extruded pieces of soft plastic tubing are pinched off and welded to the bottom of a die (Fig 9.27). Air at high pressure is blown into the soft tube causing it to *expand* and take the *shape of the mould*. The moulded object hardens on cooling and is removed by separating the two halves of the die.

8.6 Compression moulding (Formpressen)

This process is used for the moulding of objects made from *thermosetting plastics*. An appropriate amount of raw material in the form of granules is first introduced into a heated mould. When the material has softened, a plunger moves down and *compresses the material* as shown in Fig 9.28. Continued heating and pressure lead to the formation of a hard object, which is finally ejected from the mould. Objects made from thermosetting plastics remain *permanently hard* and cannot be softened and moulded again.

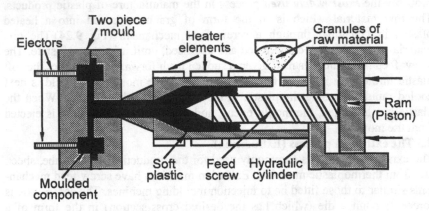

Fig 9.24 Diagram of an injection moulding machine

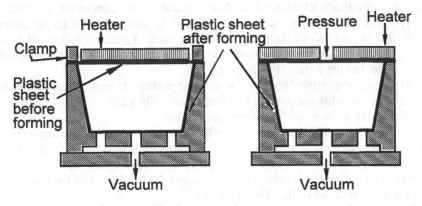

Fig 9.25 Vacuum forming   Fig 9.26 Pressure assisted vacuum forming

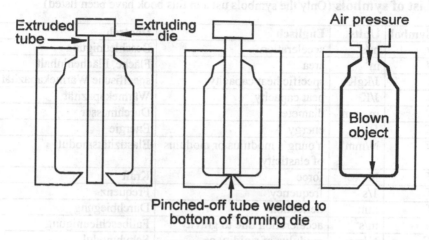

Fig 9.27 Extrusion blow moulding

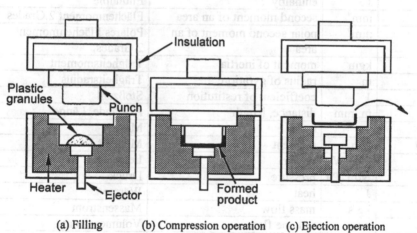

(a) Filling   (b) Compression operation   (c) Ejection operation

Fig 9.28 Compression moulding

List of symbols (Only the symbols used in this book have been listed)

Symbol	Units	Englisch	Deutsch
a	m/s^2	acceleration	Beschleunigung
A	m^2	area	Fläche, Flächeninhalt
c	J/kgK	specific heat capacity	spezifische Wärmekapazität
C	J/K	heat capacity	Wärmekapazität
d	m, mm	diameter	Durchmesser
E	J	energy	Energie
E	N/mm^2	Young's modulus or modulus of elasticity	Elastizitätsmodul
F	N	force	Kraft
f	1/s	frequency	Frequenz
f	mm	deflection	Durchbiegung
g	m/s^2	acceleration due to gravity	Fallbeschleunigung
G	N/mm^2	modulus of rigidity or shear modulus	Schubmodul
h	m	height	Höhe (allgemein)
H	J	enthalpy	Enthalpie
I	mm^4	second moment of an area	Flächenmoment 2.Grades
I_0	mm^4	polar second moment of an area	Polares Flächenmoment 2.Grades
J	kgm^2	moment of inertia	Trägheitsmoment
k	m	radius of gyration	Trägheitsradius
k	1	coefficient of restitution	Stoßzahl
l	m, mm	distance, length	Abstände, Länge
m	kg	mass	Masse
M	Nm	moment	Moment
P	W, kW	power	Leistung
P	N/m^2,Pa	pressure	Druck
Q	J	heat	Wärme
q_m	kg/s	mass flow	Massenstrom
q_v	m^3/s	volume flow	Volumenstrom
r	m,mm	radius, radius of gyration	Radius, Trägheitsradius
R_i	J/kgK	special gas constant	spezielle Gaskonstante
R	J/mol.K	universal gas constant	universelle Gaskonstante
s	m, cm, mm	displacement (vector) distance (scalar)	Weglänge
S	mm^3	elastic section modulus	Widerstandsmoment W
S	J/K	entropy	Entropie
t	s	time	Zeit
T	s	period	Periodendauer

T	Nm	twisting torque	Torsionsmoment
T	K	Kelvin temperature	Kelvin-Temperatur
u	m/s	initial velocity	Anfangsgeschwindigkeit
U	J	internal energy	Innere Energie
v	m/s	velocity	Geschwindigkeit
v	m²/s	kinematic viscosity	Kinematische Viskosität
v	1	Poisson's ratio	Poisson-Zahl
v	1	safety factor	Sicherheit
V	mm³	volume	Volumen
w	m/s	average velocity of fluid flow	Strömungsgeschwindigkeit
W	J	work	Arbeit
α	1	coefficient of contraction	Kontraktionszahl
α	1/K	coefficient of linear expansion	Längenausdehnungs-koeffizient
β	1/K	coefficient of areal expansion	Flächenausdehnungs-koeffizient
γ	1/K	coefficient of volume expansion	Volumenausdehnungs-koeffizient
γ	1	shear strain	Schubverformung
ε	1	longitudinal strain	Dehnung
ζ	1	resistance number for bends and valves in pipes	Widerstandszahl für Krümmer und Ventile
η	1	efficiency	Wirkungsgrad
η	Ns/m²	dynamic viscosity	Dynamische Viskosität
θ	rad	angle of rotation	Drehwinkel
θ	°, rad	angle	Winkel
θ	°C	Celsius temperature	Celsius-Temperatur
κ	1	isentropic exponent	Isentropenexponent
λ	1	tube flow resistance coefficient	Widerstandszahl für Rohrleitungen
μ	1	coefficient of friction	Reibungszahl (Reibzahl)
μ	1	coefficient of discharge	Ausflußzahl
ρ	°	angle of friction	Reibungswinkel
ρ	kg/m³	density	Dichte
ρ	mm	radius of curvature	Krümmungsradius
σ	N/mm²	normal stress	Normalspannung
τ	N/mm²	shear stress	Tangentialspannung
φ	1	coefficient of velocity	Geschwindigkeitszahl
φ	°, rad	angle of twist	Verdrehwinkel
ω	rad/s	angular velocity	Winkelgeschwindigkeit

Vocabulary 1
Englisch/Deutsch

Englisch	Deutsch
ability	Fähigkeit
abrasion	Abrieb, Abnutzung
abrasive particles	Schleifkörner
abrasive wheels	Schleifkörper
abrupt	plötzlich
absorb v	aufnehmen, absorbieren
absorptivity	Absorptionskraft
acceleration	Beschleunigung
accessible	zugänglich
achieve v	ausführen, erzielen
acquire v	erwerben, erlangen
act v	wirken
action	Wirkung
actuator	Stellglied
adaptive control	Maximierung der Leistung
additives	Zusatzstoffe
adhesion	Anhänglichkeit
adhesives	Klebstoffe
adiabatic	adiabate
adjacent	angrenzend, neben
adjust v	regulieren
adjusting ring or set collar	Stellringe
advantage	Vorteil
adverse	entgegenwirkend
align v	anordnen
alignment	Anordnung, Anpassung
allowable stress	zulässige Spannung
alloy	Legierung
angle	Winkel
angle milling cutters	Winkel-Stirnfräser
angular impulse	Drehimpuls
angular velocity	Winkelgeschwindigkeit
anneal v	weichglühen
apparent weight	scheinbares Gewicht
appear v	erschneinen
appearance	Aussehen
appliance	Gerät, Apparat
application	Anwendung
apply v	anwenden
approach v	sich nähern

Englisch	Deutsch
appropriate	passend, geignet
arbitrary	beliebig
arbor	Fräserdorn
arc	Bogen, Lichtbogen
area	Oberfläche
armature	Anker
arrangement	Ordnung, Einrichtung
assembly	Montageablauf
assign v	anweisen, zuteilen
associate v	verbinden, angliedern
assume v	annehmen
assumption	Annahme
at right angles	rechtwinklig
attachments	Zusatzgeräte
attract v	anziehen
automatic lathes	Drehautomaten
average value	Durchschnittswert
avoid v	vermeiden
axial locking devices	Wellensicherungen
axis	Achse
backlash	Flankenspiel
ball bearings	Kugellager
barrel finish v	trommelpolieren
basic size	Nennmaß
bead v	bördeln, falzen
beam	Träger, Tragbalken
bearing	Lager
bearings with rolling elements	Wälzlager
behave v	sich benehmen
behaviour	Verhalten, Benehmen
belt drive	Riemengetriebe
bend v	biegen
bending process	Biegeumformen
bending moment	Biegemoment
bevel gears	Kegelräder
billet	Knüppel
binder	Bindemittel
blank v	stanzen
blast furnace	Hochofen
blow	Stoß
bolts (with nuts)	Schrauben (mit Muttern)
boring bars	Bohrstangen
boundary condition	Randbedingung

brass	Messing	chucks	Futter
braze v	hartlöten	circuit	Schaltung
or hard solder v		clamps	Klemme,
break	Bruch		Spannelemente
breakdown v	abbrechen	claw-type clutches	Klauenkupplungen
breaking strength	Bruchfestigkeit	clearance angle	Freiwinkel
brittleness	Sprödigkeit	clearance fit	Spielpassung
broach v	räumen	clinker	Klinker, Schlacke
buckle v	knicken	clockwise	Uhrzeigersinn
buff v	hochglanzpolieren	clutches	schaltbare
bulk deform v	massivumformen		Kupplungen
buoyancy	Auftrieb	coarse	grob
bush	Lagerbuchse	coarse grain	grobkörnig
butt joint	Stumpfnaht	coating	Überzug, Schicht
butt weld v	Abbrennstumpf-	coat v	beschichten
	schweißen	coefficient of	Stoßzahl
cam	Nocken	restitution	
cantilever	Freiträger	coke	Koks
carbon	Kohlenstoff	collets (or collet	Spannzange
carbon content	Kohlenstoffinhalt	chucks)	
carbon steel	Kohlenstoffstahl	collide v	stoßen
carburize v	aufkohlen	collision	Stoß, Kollision
carriage	Werkzeugschlitten	column	Druckstab, Säule
carrier	Mitnehmer	column and knee	Konsolfräsmaschinen
case hardening	Einsatzstähle	type of milling	
steels		machines	
cast alloys	Gußlegierungen	combination die sets	kombinierte
cast iron	Gußeisen		Werkzeuge
catch plates	Mitnehmerscheibe	combine v	kombinieren
cause v	verursachen	commutator	Kommutator
cavity	Hohlraum	compact	kompakt, fest
centre drill	Zentrierbohrer	compare v	vergleichen
centre of gravity	Schwerpunkt	compel v	zwingen, nötigen
centre of mass	Massenmittelpunkt	compensate v	entschädigen,
centreless grinding	Spitzenlosschleifen	component	Bestandteil
(process)		composed of v	bestehen aus
centrifugal casting	Schleudergießen	composite materials	Verbundwerkstoffe
(process)		compound die sets	Gesamtschneid-
centroid	Flächenschwerpunkt		werkzeuge
chain drive	Kettengetriebe	compound table	Rechtecktisch
chamber	Kammer	compressible	komprimierbar
change of phase	Änderung des	compression	Verdichtung
	Aggregatzustandes	compression	formpressen
change of state	Zustandsänderung	mould v	
channel	Rille	compression stress	Druckspannung
charcoal	Holzkohle	compressor	Kompressor
chemical analysis	chemische Prüfungen	conclusion	Endergebnis
chemical reaction	chemische Reaktion	concrete	Beton
chisel	Meißel	condensation	Kondensation
choose v	wählen		

condition	Bedingung, Voraussetzung	dead centre	Zentrierspitze
conically	kegelförmig	decelerate v	verzögern
connecting rod	Pleuelstange	decorative	dekorativ
consider v	nachdenken, überlegen	deep draw v	tiefziehen
		defect	Mangel, Defekt
consideration	Erwägung	deficiency	Mangel, Schwäche
contact	Berührung	deflection	Durchbiegung
contain v	enthalten	deflection angle	Neigungswinkel
contamination	Verunreinigung	deform v	verformen
content	Inhalt	deformation	Verformung
continuity equation	Kontinuitäts- gleichung	degree of freedom	Freiheitsgrad
		density	Dichte
continuous	ununterbrochen	depend (on) v	abhängen (von)
continuously variable speed drives	stufenlose Getriebe	deposit v	ablagern
		deprive v	entziehen
		depth	Tiefe
drives		description	Beschreibung
contour	Kontur	design v	entwerfen
conventional	herkömmlich	desirable	wünschenswert
conversion	Umwandlung	destroy v	zerstören
conveyor	Förderband	deteriorate v	sich verschlechtern
coolant	Kühlflüssigkeit	determination	Entschlossenheit
correspond (to) v	entsprechen	determine v	bestimmen
corrosion	Korrosion	detrimental	schädlich
corrosion resistant steels	korrosionsbeständige Stähle	development	Entwicklung
		deviation	Abweichung
cost-effective	kostengünstig	diameter	Durchmesser
cotter pins	Splint	diathermic	diatherm
counterboring (process)	zylindrische Senkung	die	Schneideisen
		differ (from) v	abweichen (von)
countersinking (process)	Kegelsenkung	dimension	Dimension
		direction	Richtung
countersunk screw	Senkschraube	discontinuity	Unterbrechung
couple	Kräftepaar	disengage v	loskuppeln, befreien
couplings	Kupplungen	disperse v	zerstreuen
crack	Riß, Schlitz	displace v	verdrängen
crankshaft	Kurbelwelle	displacement (vector)	geradlinige Weglänge,
create v	erzeugen		Verschiebung
critical load	kritische Last	(compare with	
critical value	kritische Zahl	distance (scalar))	
cross-section	Durchschnitt	dissipate v	zerstreuen, verwenden
curl v	rollbiegen		
curvature	Krümmung	distance (scalar)	Weglänge
cut	Messerschneiden	distortion	Verformung
cycle	Kreislauf, Zyklus	distribute v	verteilen
cyclic process	Kreisprozess	disturbance	Störung
cylinder	Zylinder	dowel pin	Dübel
cylinder block	Zylinderblock	down mill v	gleichlauffräsen
damage v	schaden, beschädigen	draw v	ziehen
damping	Dämpfung	drill v	bohren

drive	Getriebe	evaluate *v*	berechnen, bewerten
drive shafts	Getriebewellen	evaporate *v*	verdampfen
ductile	dehnbar, biegbar	evidence	Aussage, Zeugnis
dye	Farbstoff	exceed *v*	überschreiten
ease	Leichtigkeit	excitation winding	Erregerwicklung
economical	wirtschaftlich	exhaust	Auspuff
eddy current	Wirbelstrom	expansion	Ausdehnung
effect	Wirkung	experience	Erfahrung
effective	wirksam, erfolgreich	express *v*	ausdrücken, äußern
efficiency	Wirkungsgrad	expression	Ausdruck, Formel
eject *v*	hinauswerfen	extension	Dehnung,
elastic	elastisch		Verlängerung
elastic section modulus	Widerstandsmoment	extensive	ausführlich, umfassend
elastic curve	Biegelinie	extent	Bereich, Strecke
elastomers	Elastomere	external	äußerlich
electric arc welding	Metall-Lichtbogenschweißen	external broaching (process)	Außenräumen
electron beam microanalysis	Elektronenstrahl-analyse	external cylindrical grinding (process)	Außenrundschleifen
electron beam welding	Elektronenstrahl-schweißen	extract *v*	ausziehen
electroplate *v*	galvanisieren	extrude *v*	strangpressen, verdrängen
element	Element	extrusion blow moulding process	Druckumformen
elementary forces	Teilkräfte		
eliminate *v*	beseitigen, entfernen	face mill *v*	stirnplanfräsen
elongation	Ausdehnung,	face plates	Planscheiben
emboss *v*	einprägen	facilitate *v*	erleichtern
emission	Ausstrahlung, Emission	fastener	Befestigung
		fastening device	Befestigungsgerät
empirical	empirisch	fastening pins	Befestigungsstifte
enamel	Email	fatigue strength	Dauerfestigkeit
enclose *v*	einschließen	fatigue tests	Dauerfestigkeits-prüfung
end mill *v*	stirnumfangs-planfräsen		
		feature	Besonderheit
engage *v*	betätigen, einkuppeln	feed	Vorschub
engine	mechanischer Motor (Das Wort engine ist nicht für elektrische Motoren benutzt)	feedback	Rückkopplung
		felt	Filz
		fibre-reinforced materials	faserverstärkte Verbundwerkstoffe
enlarge *v*	vergrößern	fibre	Faser
ensure *v*	sichern	file	Feile
envelope	Decke, Hülle	filler metal	Schweißstab
environment	Umgebung	fillers	Füllstoffe
equal	gleich	fillet joint	Kehlnaht
equation of state	Zustandsgleichung	final state	Endzustand
equilibrium	Gleichgewicht	fine grain	feinkörnig
equivalent	gleichwertig	fine grained welding steels	schweißgeeignete Feinkornstähle
erosion	Erosion, Abnutzung		
etch *v*	ätzen, kupferstechen		

first moment of an	Flächenmoment	glass	Glas
area	1. Grades	globules	Kügelchen
fits	Passungen	glossy	glänzend
fixed axis	feste Achse	gradient	Neigung, Gradient
fixtures	Vorrichtungen	gradual	allmählich
flakes	Flocke	grain structure	Gefüge
flame harden v	Flammenhärten	grains, crystallites	Körner, Kristallite
flange	Flansch	granules	Körnchen
flat belts	Flachriemen	graphically	graphisch
flexible shafts	biegsame Wellen	grind v	schleifen
float v	schwimmen	grinding wheel	Schleifstein
flux	Flußmittel	grip	greifen
foam	Schaum	gripper	Greifer
follower rest	mitlaufender	groove	Nut, Rille
	Setzstock	grooved pins	Kerbstifte
follow-on die set	Folgeschneid-	guide v	führen, steuern
	werkzeuge	guideways	Führungen
force	Kraft	gypsum	Gips
forge v	schmieden	hack saw	Bügelsäge
form	Form	hard cast iron	Hartguß
form v	formen, bilden	hard wearing	verschleißfest
form cutter	Profilfräser	hardness	Härte
form grind v	profilschleifen	hardness tests	Härteprüfungen
former	Former, Gestalter	headstock	Spindelstock
formula	Formel	heat capacity	Wärmekapazität
four jaw chuck	Vierbackenfutter	heat engine	Wärmekraftmaschine
fraction	Bruch, Bruchteil	heat resistant	wärmebeständig
fracture	Bruch	heat resistant steels	warmfeste Stähle
fragment	Splitter, Bruchstück	heat treatment	Wärmebehandlung
free cutting steels	Automatenstähle	height	Höhe
freeze v	frieren	helical gears	Stirnräder mit
friction	Reibung		Schrägverzahnung
friction clutches	Reibungskupplung	hemispherical	hemisphärisch
fuel	Brennstoff	high speed steels	Schnellarbeitsstähle
fundamental	Grundabmaß	hole basis system	Passungssystem
deviation			Einheitsbohrung
fundamental	Grundtoleranzgrade	homogeneous	homogen, gleichartig
tolerance grades		hot-dip coating	Schmelztauch-
fusion	Schmelzen		überzüge
fusion weld v	schmelzschweißen	identical	identisch
gap	Spalt, Öffnung	idler	Leerlaufrolle
gasket	Dichtung	ignition	Zündung
gear box	Zahnrad-	illustrate v	erläutern, darstellen
	stufengetriebe,	impact	Stoß
	Zahnrad-	impact test	Kerbschlag-
	schaltgetriebe		biegeversuch
gears	Zahnräder	impair v	schwächen
general purpose	allgemeine Stähle	imperfection	Mangel, Fehler
steels		improve v	verbessern
generate v	erzeugen	impulse	Kraftstoß

impure	unrein, gemischt	join v	verbinden
in accordance with	übereinstimmen mit	jointed shafts	Gelenkwellen
inclination	Neigung	key	Schlüssel
inclined plane	schiefe Ebene	key connections	Paßfeder
indentation	Einschnitt		Verbindungen
independent	unabhängig	keyway	Keilnut
indication	Anzeige, Andeutung	kinetic energy	kinetische Energie
indispensable	unentbehrlich	knurl v	randeln
individual	einzel	lack	Mangel
individually	einzeln	laminated materials	Schichtverbund-
induction	Induktion		werkstoffe
induction harden v	induktionshärten	lamination	Schichtung
inert gas weld v	schutzgasschweißen	lance v	ausschneiden
inertia	Trägheit	lap joint	Überlappnaht
infinitesimal	unendlich kleine	laser cut v	laserschneiden
influence	Einfluß	latent heat of fusion	Schmelzwärme
ingot	Gußblock	latent heat of	Verdampfungswärme
ingredient	Bestandteil	vapourization	
initial state	Anfangszustand	lateral	seitlich
initially	am Anfang	lathe	Drehmaschine
injection mould v	Spritzgießen	lathe centres	Zentrierspitze
input	Eingang	lathe tool	Drehmeißel
insert v	einsetzen	layer	Schicht
inspect v	prüfen	lead screw	Leitspindel
installation	Anlage	leakproof	lecksicher
instruction	Anweisung, Befehl	lengthwise	der Länge nach
insulate v	isolieren	lift v	aufheben
intensity	Intensität, Stärke	limestone	Kalkstein
interchangeability	Auswechselbarkeit	limit	Grenze, Grenzwert
interference fit	Übermaßpassung	limit gauges	Grenzlehren
intermittent	mit Unterbrechungen	limits	Grenzmaße
internal broaching	Innenräumen	line of action (of a	Wirklinie der Kraft
(process)		force)	
internal cylindrical	Innenrundschleifen	liquid lubricants	flüssige
grinding (process)			Schmierstoffe
internal energy	innere Energie	live centre	mitlaufende
interpolator	Interpolator		Zentrierspitze
intersection	Schnittpunkt	load	Last, Belastung
introduce v	einführen	load capacity	Belastbarkeit
inverse	umgekehrt	locating pins	Paßstifte
investment casting	Feingießen	location	Lage, Stelle
process		locking devices	Losdrehsicherunge
involve v	verwickeln	locking discs	Sicherungsscheibe
irregularity	Unregelmäßigkeit	locking rings	Sicherungsringe
irreversible	nicht umkehrbar	longitudinal strain	Dehnung
isolate v	isolieren	loosen v	lockern, auflockern
isotherm	Isotherme	losses	Verluste
isotropy	Isotropie	low temperature	kaltzäher Stahl
jaw	Backe	steel	
jig	Vorrichtung	lower deviation	unteres Abmaß

lower limit	Mindestmaß	modulus of	Elastizitätsmodul
lubricants	Schmierstoffe	elasticity	
lubricating greases	Schmierfette	modulus of rigidity	Schubmodul
lubrication	Schmierung	molten	geschmolzen
lubrication free	wartungsfreie	moment	Moment
bearing	Lager	moment of a couple	Drehmoment
lustrous	glänzend	moment of inertia	Trägheitsmoment
machine frame	Maschinengestell	momentum	Impuls
machining	Bearbeitungszugabe	monitor *v*	überwachen
allowance		motion	Bewegung
macromolecules	Makromoleküle	mould	Gießform
magazine	Magazine	mounting	Befestigungsschelle,
(for tools)	(für Werkzeuge)		Gestell
magnetic chuck	Magnetfutter	multipoint tool	mehrschneidiges
magnetic particle	magnetische		Werkzeug
tests	Reißprüfungen	necessary	nötig
magnitude	Größe	needle bearings	Nadellager
maintain *v*	beibehalten,	needle-shaped	nadelartig
	instandhalten	neglect *v*	vernachlässigen
malleable	hämmerbar,	neutral axis	neutrale Achse
	geschmeidig	neutral surface	neutrale Schicht
malleable cast iron	Temperguß	nibble *v*	knabberschneiden
mallet	Holzhammer	nitride *v*	nitrieren
mandrel	Drehdorn, Spandorn	nitriding steels	Nitrierstähle
manometer	Manometer	nodular cast iron	Gußeisen mit
manufacture *v*	fertigen, herstellen		Kugelgraphit
mass	Masse	nonferrous metals	Nichteisenmetalle
matter	Material, Substanz	normalizing	Normalglühen
measurement	Messung	(process)	
measure *v*	messen	notch	Kerbe, Einschnitt
melt *v*	schmelzen	oblique	schräg
melting point	Schmelzpunkt	obstacle	festes Hindernis
membrane	Membran	offset	Versetzung
metacentre	Metazentrum	omit *v*	auslassen, weglassen
metal chips	Späne	opaque	undurchsichtig
metal slitting saw	Schlitzfräser	opposite	entgegengesetzt
metallizing (or	Metall-	optical spectrum	optische
thermal spraying)	Spritzüberzüge	analysis	Spektralanalyse
metallographic	metallographische	ore	Erz
analysis	Untersuchungen	oscillate *v*	schwingen
method of sections	Schnittverfahren	outflow	Ausfluß
mild steel	Baustahl	output	Ausgang
milling cutters	Fräswerkzeuge	overhanging arm	Gegenhalter
milling machine	Fräsmaschine	overlap *v*	überlappen
misalignment	Fluchtfehler	overload *v*	überbelasten
mixture	Mischung	overload protection	Abscherstifte
models	Muster	pins	
modification	Abänderung,	overturn *v*	umkippen
	Modifikation	oxidation	Oxydierung
modify *v*	abändern		

oxy-acetylene	Gasschmelz-	polygon	Vieleck, Polygon
welding	schweißen	porous	porös
oxygen	Sauerstoff	possess v	besitzen
oxygen cutting	autogenes	possible	möglich
(process)	Brennschneiden	potential energy	potentielle Energie
paints	Lacke	power	Leistung
part	Teil	power consumption	Wirkleistung
particle-reinforced	teilchenverstärkte	predict v	vorhersagen
materials	Verbundwerkstoffe	preparation	Vorbehandlung
parting-off process	Abtrennung	preserve v	bewahren
path	Weg	press	drücken, pressen
path measurement	Wegmessung	press tools	Schneidwerkzeuge
pattern (for casting)	Modell	pressure	Druck
pendulum	Pendel	pressure pad	Preßplatte
penetrate v	eindringen	pressure resistance	Widerstands-
perforate v	perforieren	welding process	preßschweißen
perform v	leisten, ausführen	prestressed concrete	Spannbeton
performance	Leistung	prevent v	verhindern
period	Periode	previously	vorher
periodically	regelmäßig	profile	Profil, Kontur
peripheral milling	Umfangs-Planfräsen	project v	werfen
(process)		property	Eigenschaft
periphery	Umkreis, Peripherie,	protect v	schützen
	Rand, Grenze	provide v	zur Verfügung stellen
perpendicular	rechtwinklig	pulley	Rolle
physical quantity	physikalische Größe	punch	Stanzwerkzeug
piece part	Werkstück	punch clearance	Schneidspalt
(or workpiece)		purpose	Zweck, Absicht
pierce v	lochen	quantity	Menge, Anzahl
pig iron	Roheisen	quench v	abschrecken
pigments	Pigmente, Farbstoffe	quenching and	Vergütungsstähle
pilot	Suchstift	tempering steels	
pin connections	Stiftverbindungen	race (in a bearing)	Laufring
pin-ended	gelenkig gelagert	rack and pinion	Zahnstangengetriebe
pins	Stifte	radial drilling	Schwenkbohr-
pitch	Teilung	machines	maschine
plain bearing	Gleitlager	radian	rad (Radiant)
plain milling cutter	Walzenfräser	radiation	Strahlung
plane	Ebene	radioactive	radioaktiv
plasma cutting	Plasmaschneiden	radius of curvature	Krümmungsradius
(process)		radius of gyration	Trägheitsradius
plasticity	Plastizität	rake angle	Spanwinkel
plastics	Kunststoffe	rapid traverse	positionieren im
plausibility	Glaubwürdigkeit		Eilgang
plunger	Druckkolben	ratio	Verhältnis
plywood	Sperrholz	reaction	Rückwirkung
point of inflection	Inflexionspunkt	ream v	reiben
polar second	polares Flächen-	recess	Nische, Ausschnitt
moment of an area	moment 2.Grades	reciprocal	Kehrwert
polish v	polieren		

reciprocating motion	pendelnde Bewegung	roll v	rollen
		roll pins	Spannstifte
rectangular	rechtwinklig	roller bearings	Rollenlager
recycling	Wiederverwertung	rolling resistance	Rollreibung
reduce v	vermindern	rotate v	rotieren, sich drehen
reference points	Bezugspunkte	rotation	Rotation
refine v	raffinieren	rotational motion	Drehbewegung
refractory	feuerfest	rotor	Rotor, Drehteil
refractory materials	feuerfester Baustoff	rubber	Gummi
refrigerator	Kältemaschine	rust	Rost
regain v	wiedergewinnen	safety factor	Sicherheitsfaktor
regardless	ungeachtet	sample	Muster, Exemplar
regulating wheel	Regelscheibe	sand core	Kern
regulator	Regler	satisfactory	zufriedenstellend
reinforced concrete	Stahlbeton	saw	Säge
reject v	ablehnen	scale (formed on a	Schuppe,
relationship	Beziehung,	metal surface)	Metalloxydschicht
	Verhältnis	scrap	Schrott
release v	entlassen	scratch v	kratzen
reliability	Zuverlässigkeit	screw (see also bolt)	Schraube
remain v	bleiben	screw joints	Schrauben-
remove v	entfernen, beseitigen		verbindungen
repeat v	wiederholen	screw thread	Gewinde
replace v	ersetzen	seam welding	Rollennahtschweißen
represent v	darstellen	(process)	
representation	Darstellung	seamless tube	nahtlosgezogenes
require v	erfordern, brauchen		Rohr
reservoir	Speicher, Behälter	second moment of	Flächenmoment
resin	Harz	an area	2.Grades
resistance	Widerstand	section	Schnitt, Sektion
resolution of forces	Zerlegung von	select v	auswählen
	Kräften	self-centering chuck	Dreibackenbohrfutter
respond v	reagieren, antworten	sense of rotation	Sinn der Rotation
restore v	restaurieren	separate v	trennen
resultant	Resultierende	sequence	Reihenfolge
retain v	behalten	set screws	Stellschrauben
reveal v	offenbaren, enthüllen	set-up	Aufbau, Anlage
reverse v	umkehren, umdrehen	shade v	schraffieren
reverse	Gegenteil	shaft	Maschinenwelle
reversible	umkehrbar	shaft basis system	Passungssystem
rhombus	Rhombus		Einheitswelle
ribbed V-belts	mehrrippen	shaft to hub	Welle-Nabe
	Keilriemen	connections	Verbindungen
rigid body	starrer Körper	shank	Bohrerschaft
rigid couplings	starre Kupplungen	shape	Form, Gestalt
rigid shafts	starre Wellen	sharpen v	schärfen
rigidity	Festigkeit, Stabilität	shear modulus	Schubmodul
riveted joints	Nietverbindungen	shear strain	Schubverformung
rivets	Niete	shear stress	Schubspannung
rod	Rundstab	shielded arc welding	Pulver-Schweißen

shear v	scheren	spring steels	Federstähle
shot or grit blasting (process)	Körnchenblasen	spring washer	Unterlegscheibe
		sprocket wheel	Kettenrad
shot peening (process)	Verfestigungs-strahlen	spur gears	Stirnräder mit Geradverzahnung
shrink v	schrumpfen	stability	Stabilität
side milling cutter	Walzenstirnfräser	stable	stabil
similar	ähnlich	stainless steels	nichtrostende Stähle
simplify v	vereinfachen	state v	aussagen, ausstellen
single load	Einzellast	state	Zustand
sink v	sinken	state of rest	Zustand der Ruhe
sinter v	sintern	static friction	Haftreibung
sine function,	Sinusfunktion	stator	Stator
sinusoidal function		steady rest	feststehende
skill	Geschick(lichkeit)		Setzstöcke
slab	Platte	steel castings	Stahlguß
slag	Schlacke	steel for electrical	Stähle für elektrische
slenderness ratio	Schlankheitsgrad	machines	Maschinen
sliding friction	Gleitreibung	stiffness	Steifheit
sliding pillars	Führungssäule	straight	gerade
slit	Schlitz	straight line	gerade Linie
slit v	abschneiden	strain	Verformung
slope	Neigung	strain harden v	kalthärten
slot	Spalte, Nut	strand	Strang, Draht
	Schlitz	stream filaments	Stromfaden
snap rings	Springringe	stream tubes	Stromröhre
soft solder v	weichlöten	streamline flow	laminare Strömung
solder v	löten	streamlines	Stromlinien
solid lubricants	Festschmierfette	stress	Spannung
solution	Lösung	stress relieve	Spannungsarmglühen
spanner	Schraubenschlüssel	stretch v	strecken
	Steckschlüssell	stripper plate	Abstreifer
specific heat	spezifische	submerged arc	Unter-Pulver-
capacity	Wärmekapazität	weld v	Schweißen
specification	Spezifizierung	subsequent	folgend, nachträglich
specimen	Exemplar, Muster	substitute v	ersetzen
speed (scalar) (compare velocity)	Geschwindigkeit	substitute	Ersatz
		substrate	Unterschicht
speed of rotation (in revolutions per minute (rpm))	Drehzahl	successive	folgend, hintereinander
		suck in v	einsaugen
spin v	drücken, spinnen	sufficient	genug, ausreichend
spiral pins	Spiral- Spannstifte	suitable	geeignet, passend
spline	Keil	summary	Kurzfassung
splined connections	Profilwellen Verbindungen	superfinish v	kurzhubhonen
		support v	tragen, unterstützen
split (or slit) collar	geteilte Nabe	support	Träger, Stütze
spot face v	planansenken	surface	Oberfläche
spot weld v	punktschweißen	surface finish	Oberflächen-beschaffenheit
spray v	spritzen		

surface grind v	planschleifen
surface harden v	härten von
	Oberflächenschichten
surround v	umgeben
surroundings	Umgebung
swing v	schwingen
switch v	einschalten
swivel v	drehen, schwenken
tailstock	Reitstock
taper	Verjüngung
taper connection	kegliger Preßverband
taper pins	Kegelstifte
taper shank	Kegelschaft
taper roller bearings	Kegelrollenlager
tap v	gewindebohren
tap	Gewindebohrer
target value	Sollwert
task	Aufgabe
tear v	zerreißen
temper v	anlassen
tend	neigen, streben
tendency	Tendenz, Richtung
tensile stress	Zugspannung
term	fachlicher Ausdruck, Bezeichnung
thermal cutting (process)	thermisches Trenn-verfahren
thermoform	Warmumformen
thermoplastics	Thermoplaste
thermosetting plastics	Duroplaste
thinner	Verdünner
thread inserts	Gewindeeinsätze
three jaw self-centering (or universal) chuck	Dreibackenfutter
thrust (or axial force)	Axialkraft
tighten v	befestigen
tip	Spitze
tolerance	Toleranz
tool holder	Werkzeughalter, Meißelhalter
tool length compensation	Werkzeuglängen-korrektur
tool nose compensation	Schneidenradius-Korrektur
tool path compensation	Werkzeugbahn-korrektur

tool steels	Werkzeugstähle
tools	Werkzeuge
toothed belt drives	Zahnriemengetriebe
torque	Drehmoment
torsion	Verdrehung, Torsion
toughness	Zähigkeit
transfer v	übertragen
transfer of heat	Wärmeübertragung
transform v	verwandeln
transformer	Transformator
transition	Übergang
transition fit	Übergangspassung
transmission	Übertragung
transverse forces	Querkräfte
transverse loading	Querkraftbiegung
transverse section	Querschnitt
treat v	behandeln
triangle	Dreieck
trim v	beschneiden, trimmen
T-slot milling cutter	T-Nutenfräser
tube	Rohr
tubular	rohrförmig
tungsten	Wolfram
tungsten electrode process	Wolfram-Inertgas-Schweißen
turret	Revolverkopf
twist v	drehen, umdrehen
twist drill	Spiralbohrer
ultimate tensile strength	Zugfestigkeit
unbalanced	unausgeglichen
undergo v	erleben
uniform	gleichmäßig, konstant
uniform motion	gleichförmige Bewegung
unique	einzigartig
universal joints	Gelenkkupplungen
unsaturated	ungesättigt
unstable	unsicher, labil
up milling process	Gegenlauffräsen
upper deviation	oberes Abmaß
upper limit	Höchstmaß
upthrust	Auftrieb
valid	gültig, rechtskräftig
value	Wert
vapour	Dampf, Dunst
vapourization	verdunsten, verdampfen
variation	Veränderung
varnishes	Klarlacke

V-belts	Keilriemen
velocity (vector) (compare with speed (scalar))	Geschwindigkeit
vernier caliper	Meßschieber
versatile	vielseitig
vertical deflection	Durchbiegung
vessel	Gefäß
virtually	eigentlich
viscosity	Viskosität
visible	sichtbar
wall	Wand
washer	Unterlegscheibe
water-tight	wasserdicht
wavelength	Wellenlänge
weak	schwach
wear	Verschleiß
wear resistance	Verschleißfestigkeit
wedge	Keil
weight	Gewicht
wick	Docht
width	Breite
wire	Draht
wire electrode process	Metall-Schutzgasschweißen
with reference to, with respect to	bezüglich
withstand v	aushalten
wood	Holz
work	Arbeit
work harden v	kalthärten
work rest	Werkstückauflage
working conditions	Arbeitsbedingungen
worm and worm gear	Schneckengetriebe
wrench (American English)	Schraubenschlüssel
wrought alloys	Knetlegierungen
wrought iron	Schmiedeeisen
X´ray and gamma ray tests	Röntgen- und Gammastrahlen-prüfungen
X´ray fluorescence analysis	Röntgenfluoreszenz-analyse
yield strength , yield stress, yield point	Streckgrenze
zero point	Nullpunkt

Vocabulary 2
Deutsch/Englisch

Deutsch	Englisch	Deutsch	Englisch
abändern	modify *v*	Aufgabe	task
abbrechen	breakdown *v*	aufheben	lift *v*
Abbrennstumpf-	butt weld *v*	aufkohlen	carburize *v*
schweißen		absorbieren	absorb *v*
abhängen (von)	depend(on) *v*	aufnehmen,	
ablagern	deposit *v*	Auftrieb	buoyancy,
ablehnen	reject *v*		upthrust
Abrieb, Abnutzung	abrasion	Ausdehnung	expansion
Abscherstifte	overload protection	Ausdruck, Formel	expression
	pins	ausdrücken, äußern	express *v*
abschneiden	(to) slit *v*	Ausfluß	outflow
abschrecken	quench *v*	ausführen, erzielen	achieve *v*
Absorptionsgrad	absorptivity	ausführlich	extensive
Abstreifer	stripper plate	Ausgang	output
Abtrennung	parting-off process	aushalten	withstand *v*
abweichen (von)	differ (from) *v*	auslassen, weglassen	omit *v*
Abweichung	deviation	Auspuff	exhaust
Achse	axis	Aussage	evidence
adiabate	adiabatic	aussagen, ausstellen	state *v*
ähnlich	similar	ausschneiden	lance *v*
allgemeine Stähle	general purpose steels	Aussehen	appearance
allmählich	gradual	Außenräumen	external broaching
am Anfang	initially		process
Änderung des	change of phase	Außenrundschleifen	external cylindrical
Aggregatzustandes			grinding process
Anfangszustand	initial state	äußerlich	external
angrenzend, neben	adjacent	Ausstrahlung	emission
Anhänglichkeit	adhesion	auswählen	select *v*
Anker	armature	Auswechselbarkeit	interchangeability
Anlage	installation	ausziehen	extract *v*
anlassen	temper *v*	autogenes	oxygen cutting
Annahme	assumption	Brennschneiden	process
annehmen	assume *v*	Automatenstähle	free cutting steels
anordnen	align *v*	Axialkraft	thrust (or axial force)
reagieren	respond *v*	Backe	jaw
anweisen, zuteilen	assign *v*	Baustahl	mild steel
Anweisung, Befehl	instruction	Bearbeitungs-	machining allowance
anwenden	apply *v*	zugabe	
Anwendung	application	Bedingung,	condition
Anzeige, Andeutung	indication	Voraussetzung	
anziehen	attract *v*	befestigen	tighten *v*
Arbeit	work	Befestigung	fastener
Arbeitsbedingungen	working conditions	Befestigungsgerät	fastening device
ätzen, kupferstechen	etch *v*	Befestigungsschelle,	mounting
Aufbau, Anlage	set-up	Gestell	

Befestigungsstifte	fastening pins
behalten	retain *v*
behandeln	treat *v*
beibehalten	maintain *v*
Belastbarkeit	load capacity
beliebig	arbitrary
berechnen,	evaluate *v*
bewerten	
Bereich, Strecke	extent
Berührung	contact
beschichten	coat *v*
Beschleunigung	acceleration
beschneiden	trim *v*
Beschreibung	description
beseitigen, entfernen	eliminate *v*
besitzen	possess *v*
Besonderheit	feature
Bestandteil	ingredient, component
bestehen aus	composed of *v*
bestimmen	determine *v*
betätigen,	engage *v*
einkuppeln	
Beton	concrete
bewahren	preserve *v*
Bewegung	motion
Beziehung	relationship
bezüglich	with reference to, with respect to
Bezugspunkte	reference points
Biegelinie	elastic curve
Biegemoment	bending moment
biegen	bend *v*
biegsame Wellen	flexible shafts
Bindemittel	binder
bleiben	remain *v*
Bogen, Lichtbogen	arc
bohren	drill *v*
Bohrerschaft	shank
Bohrstangen	boring bars
bördeln, falzen	bead *v*
Breite	width
Brennstoff	fuel
Bruch	break, fracture
Bruch, Bruchteil	fraction
Bruchfestigkeit	breaking strength
Bügelsäge	hack saw
chemische Prüfungen	chemical analysis
chemische Reaktion	chemical reaction

Dampf, Dunst	vapour
Dämpfung	damping
darstellen	represent *v*
Darstellung	representation
Dauerfestigkeit	fatigue strength
Dauerfestigkeits-prüfung	fatigue tests
Decke, Hülle	envelope
dehnbar, biegbar	ductile
Dehnung	longitudinal strain
dekorativ	decorative
der Länge nach	lengthwise
diatherm	diathermic
Dichte	density
Dichtung	gasket
Dimension	dimension
Docht	wick
Draht	wire
Drehautomaten	automatic lathes
Drehbewegung	rotational motion
Drehdorn, Spanndorn	mandrel
drehen, schwenken	swivel *v*
drehen, umdrehen	twist *v*
Drehimpuls	angular impulse
Drehmaschine	lathe
Drehmeißel	lathe tool
Drehmoment	moment of a couple, torque
Drehzahl	speed of rotation (in revolutions per minute (rpm))
Dreibackenfutter	three jaw self-centering (or universal) chuck
Dreieck	triangle
Druck	pressure
drücken, pressen	press *v*
drücken, spinnen	spin *v*
Druckkolben	plunger
Druckspannung	compression stress
Druckstab, Säule	column
Druckumformen	extrusion blow moulding (process)
Dübel	dowel pin
Durchbiegung	deflection
durchgehen	traverse *v*
Durchmesser	diameter
Durchschnitt	cross-section
Durchschnittswert	average value

Duroplaste	thermosetting plastics	Erregerwicklung	excitation winding
Ebene	plane	erscheinen	appear v
Eigenschaft	property	ersetzen	replace, substitute v
eigentlich	virtually	Ersatz	substitute
eindringen	penetrate v	Erwägung	consideration
einfache Zentrier-	dead centre	erwerben, erlangen	acquire v
spitze		Erz	ore
Einfluß	influence	erzeugen	create v, generate v
einführen	introduce v	Exemplar, Muster	specimen
Eingang	input	flüssige Schmier-	liquid lubricants
einprägen	emboss v	stoffe	
Einsatzstähle	case hardening steels	Fähigkeit	ability
einsaugen	suck in v	Farbstoff	dye
einschalten	switch v	Faser	fibre
einschließen	enclose v	faserverstärkte	fibre-reinforced
Einschnitt	indentation	Verbundwerkstoffe	materials
einsetzen	insert v	Federstähle	spring steels
einzeln	individual	Feile	file
Einzellast	single load	Feingießen	investment casting
einzigartig	unique		(process)
elastisch	elastic	feinkörnig	fine grain
Elastizitätsmodul	modulus of elasticity	fertigen, herstellen	manufacture v
Elastomere	elastomers	feste Achse	fixed axis
Elektronenstrahl-	electron beam	festes Hindernis	obstacle
analyse	microanalysis	Festigkeit, Stabilität	rigidity
Elektronenstrahl-	electron beam	Festschmierfette	solid lubricants
schweißen	welding	feststehender	steady rest
Element	element	Setzstock	
Email	enamel	feuerfest	refractory
empirisch	empirical	feuerfester Baustoff	refractory materials
Endergebnis	conclusion	Filz	felt
Endzustand	final state	Flächenmoment	first moment of an
entfernen, beseitigen	remove v	1. Grades	area
entgegengesetzt	opposite	Flächenmoment	second moment of an
entgegenwirkend	adverse	2. Grades	area
enthalten	contain v	Flächenschwer-	centroid
entlassen	release v	punkt	
Entschlossenheit	determination	Flachriemen	flat belts
entschädigen,	compensate v	flammhärten	flame harden v
ersetzen	replace v	Flankenspiel	backlash
entsprechen	correspond to v	Flansch	flange
entwerfen	design v	Flocke	flakes
Entwicklung	development	Fluchtfehler	misalignment
entziehen	deprive v	Flußmittel	flux
Erfahrung	experience	folgend	successive
erfordern, brauchen	require v	nachträglich	subsequent
Erhaltung	conservation	Folgeschneid-	follow-on die set
erläutern, darstellen	illustrate v	werkzeuge	
erleichtern	facilitate v	Förderband	conveyor
Erosion, Abnutzung	erosion	Form	form

Form, Gestalt	shape	Gießform	mould
Formel	formula	Gips	gypsum
formen, bilden	form v	glänzend	glossy,
Former, Gestalter	former		lustrous
formpressen	compression mould v	Glas	glass
Fräsmaschine	milling machine	Glaubwürdigkeit	plausibility
Fräswerkzeuge	milling cutters	gleich	equal
Fräserdorn	arbor	gleichförmige	uniform motion
Freiheitsgrad	degree of freedom	Bewegung	
Freiträger	cantilever	Gleichgewicht	equilibrium
Freiwinkel	clearance angle	gleichlauffräsen	down mill v
frieren	freeze v	gleichmäßig,	uniform
führen, steuern	guide v	konstant	
Führungen	guideways	gleichwertig	equivalent
Führungssäule	sliding pillars	Gleitlager	plain bearing
Füllstoffe	fillers	Gleitreibung	sliding friction
Futter	chucks	graphisch	graphically
galvanisieren	electroplate v	greifen	grip v
Gasschmelz-	oxy-acetylene	Greifer	gripper
schweißen	welding (process)	Grenze, Grenzwert	limit
Gefäß	vessel	Grenzlehren	limit gauges
Gefüge	grain structure	Grenzmaße	limits
Gegenhalter	overhanging arm	grob	coarse
Gegenlauffräsen	up milling process	grobkörnig	coarse grain
Gegenteil	reverse	Größe	magnitude
geeignet, passend	suitable	Grundabmaß	fundamental
gelenkig gelagert	pin-ended		deviation
Gelenkkupplungen	universal joints	Grundtoleranzgrade	fundamental
Gelenkwellen	jointed shafts		tolerance grades
genug, ausreichend	sufficient	gültig, rechtskräftig	valid
gerade	straight	Gummi	rubber
gerade Linie	straight line	Gußblock	ingot
geradlinige	displacement (vector)	Gußeisen	cast iron
Weglänge	(compare with	Gußeisen mit	nodular cast iron
	distance)	Kugelgraphit	
Gerät, Apparat	appliance	Gußlegierungen	cast alloys
Gesamtschneid-	compound die sets	Haftreibung	static friction
werkzeuge		hämmerbar,	malleability
Geschick(lichkeit)	skill	geschmeidig	
geschmolzen	molten	Härte	hardness
Geschwindigkeit	speed (scalar),	Härteprüfungen	hardness tests
	velocity (vector)	Hartguß	hard cast iron
geteilte Nabe	split (or slit) collar	hartlöten	braze v
Getriebe	drive		or hard solder v
Getriebewellen	drive shafts	härten von	surface harden v
Gewicht	weight	Oberflächen-	
Gewinde	screw thread	schichten	
gewindebohren	tap v	Harz	resin
Gewindebohrer	tap	hemisphärisch	hemispherical
Gewindeeinsätze	thread inserts	herkömmlich	conventional

hinauswerfen	eject v	Kerbschlag-	impact test
hochglanzpolieren	buff v	biegeversuch	
Hochleistungs-	heavy duty	Kerbstifte	grooved pins
Hochofen	blast furnace	Kern	sand core
Höchstmaß	upper limit	Kettengetriebe	chain drive
Höhe	height	Kettenrad	sprocket wheel
Hohlraum	cavity	kinetische Energie	kinetic energy
Holz	wood	Klarlack	varnish
Holzhammer	mallet	Klauenkupplungen	claw-type clutches
Holzkohle	charcoal	Klebstoffe	adhesives
homogen,	homogeneous	Klemme	clamps
gleichartig		Klinker, Schlacke	clinker
identisch	identical	knabberschneiden	nibble v
Impuls	momentum	Knetlegierungen	wrought alloys
Induktion	induction	knicken	buckle v
induktionshärten	induction harden v	Knüppel	billet
Inflexionspunkt	point of inflection	Kohlenstoff	carbon
Inhalt	content	Kohlenstoffinhalt	carbon content
innenräumen	internal broach v	Kohlenstoffstahl	carbon steel
innere Energie	internal energy	Koks	coke
Innernrundschleifen	internal cylindrical	kombinieren	combine v
	grinding (process)	kombinierte	combination die sets
instandhalten	maintain v	Werkzeuge	
Intensität, Stärke	intensity	Kommutator	commutator
Interpolator	interpolator	kompakt, fest	compact
isolieren	insulate v,	Kompressor	compressor
	isolate v	komprimierbar	compressible
Isotherme	isotherm	Kondensation	condensation
Isotropie	isotropy	Konsolfräsmachinen	column and knee type
Kalkstein	limestone		millimg machines
Kältemaschine	refrigerator	Kontinuitäts-	continuity equation
kalthärten	strain harden v,	gleichung	
	work harden v	Kontur	contour
kaltzäher Stahl	low temperature steel	Körnchen	granules
Kammer	chamber	Körnchenblasen	shot or grit blasting
kegelförmig	conically		(process)
Kegelräder	bevel gears	Körner, Kristallite	grains, crystallites
Kegelrollenlager	taper roller bearings	Korrosion	corrosion
Kegelschaft	taper shank	korrosions-	corrosion resistant
kegelsenken	countersink v	beständige Stähle	steels
Kegelstifte	taper pins	kostengünstig	cost-effective
kegliger	taper connection	Kraft	force
Preßverband		Kräftepaar	couple
Kehlnaht	fillet joint	Reibungskupplung	friction clutches
Kehrwert	reciprocal	Kraftstoß	impulse
Keil	wedge, spline	kratzen, Kratzer	scratch
Keilnut	keyway	Kreislauf, Zyklus	cycle
Keilriemen	V-belts	Kreisprozess	cyclic process
Keilsitzverbindung	taper key connection	kritische Last	critical load
Kerbe, Einschnitt	notch	kritische Zahl	critical value

Krümmung	curvature
Krümmungsradius	radius of curvature
Kügelchen	globules
Kugellager	ball bearings
Kühlflüssigkeit	coolant
Kunststoffe	plastics
Kupplungen	couplings
Kurbelwelle	crankshaft
Kurzfassung	summary
kurzhubhonen	superfinish v
Lacke	paints
Lage, Stelle	location
Lager	bearing
Lagerbuchse	(bearing) bush
laminare Strömung	streamline flow
laserschneiden	laser cutting v
Last, Belastung	load
Laufring	race (in a bearing)
lecksicher	leakproof
Leerlaufrolle	idler
Legierung	alloy
Leichtigkeit	ease
leisten, ausführen	perform v
Leistung	power
Leitspindel	lead screw
lochen	pierce v
lockern, auflockern	loosen v
Losdrehsicherung	locking devices
loskuppeln, befreien	disengage v
Lösung	solution
löten	solder v
Magazine (für	magazine
Werkzeuge)	(for tools)
Magnetfutter	magnetic chuck
magnetische	magnetic particle
Reißprüfungen	tests
Makromoleküle	macromolecules
Mangel	defect, deficiency
	imperfection
Manometer	manometer
Maschinengestell	machine frame
Maschinenwelle	shaft
Masse	mass
Massenmittelpunkt	centre of mass
massivumformen	bulk deform v
Material, Substanz	matter
Maximierung der	adaptive control
Leistung	
mechanischer Motor	engine
(Das Wort engine ist	
nicht für elektrische	
Motoren benutzt)	
Mehrrippen-	ribbed V-belts
keilriemen	
mehrschneidiges	multipoint tool
Werkzeug	
Meißel	chisel
Membran	membrane
Menge, Anzahl	quantity
messen	measure v
messerschneiden	cut v
Messing	brass
Meßschieber	vernier caliper
Messung	measurement
Metall-Lichtbogen-	electric arc welding
schweißen	process
metallographische	metallographic
Untersuchungen	analysis
Metall-Schutzgas-	wire electrode
schweißen	process
Metall-Spritz-	metallizing or
überzüge	thermal spraying
Metazentrum	metacentre
Mindestmaß	lower limit
Mischung	mixture
mit Unter-	intermittent
brechungen	
mitlaufende	live centre
Zentrierspitze	
mitlaufender	follower rest
Setzstock	
Mitnehmer	carrier
Mitnehmerscheibe	catch plates
Modell	pattern (for casting)
möglich	possible
Moment	moment
Montageablauf	assembly
Muster, Exemplar	sample
nachdenken,	consider v
überlegen	
nadelartig	needle-shaped
Nadellager	needle bearings
nahtlosgezogenes	seamless tube
Rohr	
Neigung, Steigung,	inclination,slope,
Gradient	gradient
Neigungswinkel	deflection angle
Nennmaß	basic size
neutrale Achse	neutral axis
neutrale Schicht	neutral surface

nicht umkehrbar	irreversible	Pleuelstange	connecting rod
Nichteisenmetalle	nonferrous metals	plötzlich	abrupt
nichtrostende Stähle	stainless steels	polares Flächen-	polar second moment
Niete	rivets	moment 2.Grades	of an area
Nietverbindungen	riveted joints	polieren	polish v
Nische, Ausschnitt	recess	porös	porous
nitrieren	nitride v	positionieren im	rapid traverse v
Nitrierstähle	nitriding steels	Eilgang	
Nocken	cam	potentielle Energie	potential energy
Normalglühen	Normalizing	Preßplatte	pressure pad
	(process)	Profil, Kontur	profile
nötig	necessary	Profilfräser	form cutter
Nut, Rille	groove	profilschleifen	form grind v
oberes Abmaß	upper deviation	Profilwellen-	splined connections
Oberfläche	surface,	Verbindungen	
	surface area,	prüfen	inspect v
Oberflächen-	surface finish	pulverschweißen	shielded arc weld v
beschaffenheit		punktschweißen	spot weld v
Oberflächenfein-	surface finishing	Querkraftbiegung	transverse loading
bearbeitung	processes	Querkräfte	transverse forces
offenbaren,	reveal v	Querschnitt	transverse section
enthüllen		rad (Radiant)	radian
optische	optical spectrum	radioaktiv	radioactive
Spektralanalyse	analysis	raffinieren	refine
Ordnung,	arrangement	Randbedingung	boundary condition
Einrichtung		randeln	knurl v
Oxydierung	oxidation	räumen	broach v
passend, geignet	appropriate	Rechtecktisch	compound table
Paßfeder-Verbind-	key connections	rechtwinklig	at right angles,
ungen			perpendicular,
Paßstifte	locating pins		rectangular
Passungen	fits	regelmäßig	regular, periodical
Passungssystem	hole basis system	Regelscheibe	regulating wheel
Einheitsbohrung		Regler	regulator
Passungssystem	shaft basis system	regulieren	adjust v
Einheitswelle		reiben	ream v
Pendel	pendulum	Reibung	friction
pendelnde	reciprocating motion	Reihenfolge	sequence
Bewegung		Reitstock	tailstock
perforieren	perforate v	restaurieren	restore v
Periode	period	Resultierende	resultant
physikalische Größe	physical quantity	Revolverkopf	turret
Pigmente,	pigments	Rhombus	rhombus
Farbstoffe		Richtung	direction
planansenken	spot face v	Riemengetriebe	belt drive
Planscheiben	face plates	Rille	channel
Plasmaschneiden	plasma cutting	Riß, Schlitz	crack
	process	Roheisen	pig iron
planschleifen	surface grind v	Rohr	tube
Platte	slab	rohrförmig	tubular

rollbiegen	curl v	schmelzschweißen	fusion weld v
Rolle	pulley	Schmelztauch-	hot-dip coating
rollen	roll v	überzüge	
Rollenlager	roller bearings	Schmelzwärme	latent heat of fusion
Rollennaht-	seam welding	Schmiedeeisen	wrought iron
schweißen	(process)	schmieden	forge v
Rollreibung	rolling resistance	Schmierfette	lubricating greases
Röntgen- und	X´ray and gamma ray	Schmierstoffe	lubricants
Gammastrahlen-	tests	Schmierung	lubrication
prüfungen		Schneckengetriebe	worm and worm gear
Röntgenfluoreszenz-	X´ray fluorescence	Schneideisen	die (for thread
analyse	analysis		cutting)
Rost	rust	Schneidenradius-	tool nose
Rotation	rotation	Korrektur	compensation
rotieren, sich drehen	rotate v	Schneidspalt	punch clearance
Rotor, Drehteil	rotor	Schneidwerkzeuge	press tools
Rückkopplung	feedback	Schnellarbeitsstähle	high speed steels
Rückwirkung,	reaction	Schnitt, Sektion	section
Säge	saw	Schnittpunkt	intersection
Sauerstoff	oxygen	Schnittverfahren	method of sections
schaden,	damage v	schraffieren	shade v
beschädigen		schräg	oblique
schädlich	detrimental	Schraube	screw (see also bolt)
schaltbare	clutches	Schrauben-	screw joints
Kupplungen		verbindungen	
Schaltung	circuit	Schrauben (mit	bolts (with nuts)
schärfen	sharpen v	Muttern)	
Schaum	foam	Schraubenschlüssel	spanner,
scheinbares Gewicht	apparent weight	Steckschlüssell	wrench (american)
scheren	shear v	Schrott	scrap
Schicht	layer	schrumpfen	shrink v
Schichtung	lamination	Schubmodul	modulus of rigidity
Schichtverbund-	laminated materials	Schubspannung	shear stress
werkstoffe		Schubverformung	shear strain
schiefe Ebene	inclined plane	Metalloxydschicht,	scale (formed on a
Schild, Schirm	shield	Schuppe	metal surface)
Schlacke	slag	schützen	protect v
Schlankheitsgrad	slenderness ratio	Schutzgasschweißen	inert gas welding
schleifen	grind v		process
Schleifkörner	abrasive particles	schwächen	impair v
Schleifkörper	abrasive wheels	Schweißstab	filler metal
Schleifstein	grinding wheel	schweißgeeignete	fine grained welding
Schleudergießen	centrifugal casting	Feinkornstähle	steels
	process	Schwenkbohr-	radial drilling
Schlitzfräser	metal slitting saw	maschine	machines
Schlitz	slit, slot	Schwerpunkt	centre of gravity
Schlüssel	key	schwimmen	float v
Schmelzen	fusion (process)	schwingen	oscillate v,
schmelzen	melt v		swing v
Schmelzpunkt	melting point	seitlich	lateral

Senkschraube	countersunk screw	Stahlbeton	reinforced concrete
sich benehmen	behave v	Stähle für elektri-	steel for electrical
sich nähern	approach v	sche Maschinen	machines
sich verschlechtern	deteriorate v	Stahlguß	steel castings
Sicherheitsfaktor	safety factor	stanzen	blanking process
sichern	ensure v	Stanzwerkzeug	punch
Sicherungsringe	locking rings	starre Körper	rigid body
Sicherungsscheibe	locking discs	starre Kupplungen	rigid couplings
sichtbar	visible	starre Wellen	rigid shafts
sinken	sink v	Stator	stator
Sinn der Rotation	sense of rotation	Steifheit	stiffness
sintern	sinter v	Stellglied	actuator
Sinusfunktion	sinusoidal function, sine function	Stellringe	adjusting ring or set collar
Sollwert	target value	Stellschrauben	set screws
Spalt, Öffnung	gap	Stifte	pins
Spalte, Nut	slot	Stiftverbindungen	pin connections
Späne	metal chips	stirnplanfräsen	face mill v
Spannbeton	prestressed concrete	Stirnräder mit	helical gears
Spannelemente	clamps, clamping devices	Schrägverzahnung Stirnräder mit	spur gears
Spannstifte	roll pins	Geradverzahnung	
Spannung	stress	stirnumfangs-	end mill v
Spannungsarm-	stress relieving	planfräsen	
glühen	(process)	Störung	disturbance
Spannzange	collets (or collet chucks)	Stoß	blow, impact
		Stoß, Kollision	collision
Spanwinkel	rake angle	stoßen	collide v
Speicher, Behälter	reservoir	Stoßzahl	coefficient of restitution
Sperrholz	plywood		
spezifische	specific heat capacity	Strahlung	radiation
Wärmekapazität		Strang, Draht	strand
Spezifizierung	specification	strangpressen,	extrude v
Spielpassung	clearance fit	verdrängen	
Spindel	spindle	strecken	stretch v
Spindelstock	headstock	Streckgrenze	yield strength , yield stress, yield point
Spiral-Spannstifte	spiral pins		
Spiralbohrer	twist drill	Stromfaden	stream filaments
Spitze	tip	Stromlinien	streamlines
Spitzenlosschleifen	centreless grinding process	Stromröhre	stream tubes
		Stufenlosegetriebe	continuously variable speed drives
Splint	cotter pins		
Splitter, Bruchstück	fragment	Stumpfnaht	butt joint
Springringe	snap rings	Suchstift	pilot
spritzen	spray v	Teil	part
spritzgießen	injection mould v	teilchenverstärkte	particle-reinforced
Sprödigkeit	brittleness	Verbundwerkstoffe	materials
Stab (rund)	rod	Teilkräfte	elementary forces
stabil	stable	Teilung	pitch
Stabilität	stability	Temperguß	malleable cast iron

Tendenz, Richtung	tendency	unrein, gemischt	impure
thermisches Trenn-verfahren	thermal cutting	unsicher, labil	unstable
		Unterbrechung	discontinuity
Thermoplaste	thermoplastics	unteres Abmaß	lower deviation
Tiefe	depth	Unterlegscheibe	washer, spring washer
tiefziehen	deep draw v		
T-Nutenfräser	T-slot milling cutter	Unter-Pulver-Schweißen	submerged arc weld v
Toleranz	tolerance		
tragen, unterstützen	support v	Unterschicht	substrate
Träger, Stütze	support	ununterbrochen	continuous
Tragbalken	beam	Veränderung	variation
Trägheit	inertia	verbessern	improve v
Trägheitsmoment	moment of inertia	verbinden	join v
Trägheitsradius	radius of gyration	Verbundwerkstoffe	composite materials
Transformator	transformer	verdampfen	evaporate v
trennen	separate v	Verdampfungs-wärme	latent heat of vapourization
trommelpolieren	barrel finish v		
überbelasten	overload v	Verdichtung	compression
übereinstimmen mit	in accordance with	verdrängen	displace v
Übergang	transition	Verdrehung, Torsion	torsion
Übergangspassung	transition fit		
überlappen	overlap v	Verdünner	thinner
Überlappnaht	lap joint	verdunsten	vapourize v
Übermaßpassung	interference fit	vereinfachen	simplify v
überschreiten	exceed v	Verfestigungs-strahlen	shot peening (process)
übertragen	transfer v		
Übertragung	transmission	verformen	deform v
überwachen	monitor v	Verformung	deformation, distortion
Überzug, Schicht	coating		
Uhrzeigersinn	clockwise	vergleichen	compare v
umdrehen	twist, rotate	vergrößern	enlarge v
Umfangs-Planfräsen	peripheral milling (process)	Vergütungsstähle	quenching and tempering steels
umgeben	surround v	Verhalten	behaviour
Umgebung	environment, surroundings	Verhältnis	ratio
		verhindern	prevent v
umgekehrt	inverse	Verjüngung	taper
umkehrbar	reversible	Verlängerung	elongation
umkehren	reverse v	Verluste	losses
umkippen	overturn v	vermeiden	avoid v
Umkreis, Peripherie	periphery	vermindern	reduce v
Umwandlung	conversion	vernachlässigen	neglect v
unabhängig	independent	Verschleiß	wear
unausgeglichen	unbalanced	verschleißfest	hard wearing
undurchsichtig	opaque	Verschleißfestigkeit	wear resistance
unendlich klein	infinitesimal	Versetzung	offset
unentbehrlich	indispensable	verteilen	distribute v
ungeachtet	regardless	Verunreinigung	contamination
ungesättigt	unsaturated	verursachen	cause v
Unregelmäßigkeit	irregularity	verwandeln	transform v

verwerfen	reject *v*		korrektur	compensation
verwickeln	involve *v*		Werkzeugstähle	tool steels
verzögern	decelerate *v*		Wert	value
Vieleck, Polygon	polygon		Widerstand	Resistance
vielseitig	versatile		Widerstands-	elastic section
Vierbackenfutter	four jaw chuck		moment	modulus
Viskosität	viscosity		Widerstands-	pressure resistance
Vorbehandlung	preparation		preßschweißen	welding process
vorher	previously		wiedergewinnen	regain *v*
vorhersagen	predict *v*		wiederholen	repeat *v*
Vorrichtungen	jigs, fixtures		Wiederverwertung	recycling process
Vorschub	feed		Winkel	angle
Vorteil	advantage		Winkel-	angular velocity
wählen	choose *v*		geschwindigkeit	
Walzenfräser	plain milling cutter		Winkel-Stirnfräser	angle milling cutters
Walzenstirnfräser	side milling cutter		Wirbelstrom	eddy current
Wälzlager	bearings with rolling		Wirkleistung	power consumption
	elements		Wirklinie der Kraft	line of action (of a
Wand	wall			force)
Wärmebehandlung	heat treatment		wirksam	effective
wärmebeständig	heat resistant		Wirkung	effect
Wärmekapazität	heat capacity		Wirkungsgrad	efficiency
Wärmekraft-	heat engine		wirtschaftlich	economical
maschine			Wolfram	tungsten
Wärmeübertragung	transfer of heat		Wolfram-Inertgas-	tungsten electrode
warmfeste Stähle	heat resistant steels		Schweißen	welding process
warmumformen	thermoform *v*		wünschenswert	desirable
wartungsfreie	lubrication free		Zähigkeit	toughness
Lager	bearing		Zahnräder	gears
wasserdicht	water-tight		Zahnrad-	gear box
Weg	path		schaltgetriebe	
Weglänge	distance (scalar)		Zahnriemengetriebe	toothed belt drives
Wegmessung	path measurement		Zahnstangen-	rack and pinion
weichglühen	anneal *v*		getriebe	
weichlöten	soft solder *v*		Zentrierbohrer	centre drill
Welle-Nabe	shaft to hub		Zentrierspitze	lathe centres
Verbindungen	connections		Zerlegung von	resolution of forces
Wellenlänge	wavelength		Kräften	
Wellensicherungen	axial locking devices		zerreißen	tear *v*
werfen	project *v*		zerstören	destroy *v*
Werkstück	piece part		zerstreuen	disperse *v*
	(or workpiece)		ziehen	draw *v*
Werkstückauflage	work rest		zufriedenstellend	satisfactory
Werkzeugbahn-	tool path		zugänglich	accessible
korrektur	compensation		Zugfestigkeit	ultimate tensile
Werkzeuge	tools			strength
Werkzeughalter	tool holder		Zugspannung	tensile stress
Meißelhalter	Lathe tool holder		zulässige Spannung	allowable stress
Werkzeugschlitten	carriage		Zündung	ignition
Werkzeuglängen-	tool length			

Zusatzgeräte	attachments
Zusatzstoffe	additives
Zustand	state
Zustand der Ruhe	state of rest
Zustandsänderung	change of state
Zustandsgleichung	equation of state
Zuverlässigkeit	reliability
Zweck, Absicht	purpose
zwingen	compel *v*
Zylinder	cylinder
Zylinderblock	cylinder block
zylindrisch senken	counterbore *v*

Index

Stichwortverzeichnis

Bibiliography

This is only a small selection from the large number of books available in English. New editions of these books are published frequently and it is very difficult to keep track of these changes. For this reason, the edition number and year of publication are not given below.

1. Avallone and Baumeister : Marks´ Standard Handbook for Mechanical Engineers (McGraw-Hill)
2. Kutz: Mechanical Engineer´s Handbook (John Wiley)
3. Dubbel: Taschenbuch für den Maschinenbau (Springer)
4. Alfred Böge: Techniker Handbuch (Vieweg)
5. Rothbart: Mechanical Design Handbook (McGraw-Hill)
6. Meriam and Craig: Engineering Mechanics, Vol. I: Statics, Vol II: Dynamics (John Wiley)
7. Beer and Johnston: Vector Mechanics for Engineers: Statics, Dynamics (McGraw-Hill)
8. Beer and Johnston: Mechanics of Materials (McGraw-Hill)
9. Riley and Sturges: Statics and Mechanics of materials (John Wiley)
10. Shames: Mechanics of Fluids (McGraw-Hill)
11. Young, Munson and Okiishi: A Brief Introduction to Fluid Mechanics (John Wiley)
12. Oertel: Introduction to Fluid Mechanics (Vieweg)
13. Degarmo, Black and Kohser : Materials and Processes in Manufacturing (John Wiley)
14. Schey: Introduction to Manufacturing Processes (McGraw-Hill)
15. Wark: Thermodynamics (McGraw-Hill)
16. Moran and Shapiro: Fundamentals of Engineering Thermodynamics (John Wiley)
17. Zemansky and Dittman: Heat and Thermodynamics (McGraw-Hill)
18. Schigley, Mischke and Budynas: Mechanical Engineering Design (McGraw-Hill)
19. Collins, Staab and Busby: Design of Machine Elements and Machines (John Wiley)
20. Krar and Oswald: Technology of Machine Tools (McGraw-Hill)
21. Thyer: Computer Numerical Control of Machine Tools (Industrial Press)
22. Kief and Waters: Computer Numerical Control (McGraw-Hill)
23. Bollinger and Duffle: Computer Control of Machines and Processes (Addison-Wesley)
24. Krar and Gill: CNC Technology and Programming (McGraw-Hill)
25. Kalpakjian: Manufacturing Engineering and Technology (Addison-Wesley)
26. Kalkapjian: Manufacturing Processes for Engineering Materials (Addison-Wesley)

Appendix 1

Alternate word forms in English

English is a language that is used and spoken all over the world, with the consequence that english words can be found in many other languages. On the other hand, english has also acquired words from other languages, like for example "Guru" or "Mantra" from sanskrit, and "Angst" or "Eigenfunktion" from german. With increasing globalization, it is possible that many regional forms of english may be created in the future, similar to that of american english, which can be considered to be a regional form of english.

In this book the british kind of english has been used, and as is well known, the word forms are closer to the latin than the word forms used in american english. Readers who use american textbooks may want to know something about the difference between british and american word forms, and also about the difference between the british and american pronounciation of a given word. The difference in most cases is small, and examples are given below to illustrate a few cases where a difference does exist. Those interested in knowing more, should refer to the dictionaries mentioned below.

A few examples of differences in spelling

- Some words which end in "*ise*" like *stabilise* can also be written as *stabilize*. The word form *stabilise* is more commonly used in Britain .
- Words ending in "*our*" in british english like *vapour* and *colour* are written in american english as *vapor* and *color*.
- The word *centre* in british english is written as *center* in american english.
- In some cases the singular forms are the same, while the plural form may be different. Examples are singular forms like *radius* and *index* which are the same in british and american english. These have the plural forms *radii* and *indices* in british english (following the latin plural forms), while the american forms are *radiuses* and *indexes*.

Differences in technical terms

In most cases the British and American technical terms are the same. Only in rare cases is there an outright difference. A few cases are quoted below.

- The hand tool which is called a *spanner* (Schlüssel) in british english is termed a *wrench* in american english.
- The measuring device *gauge* (Lehre) in british english is written *gage* in american english. Both forms are pronounced in the same way.
- The measuring devices called *slip gauges* (Endmasse) in british english are called *gage blocks* in american english.
- The fuel used in automobiles called *petrol* (Benzene) in british english is called *gasoline* or *gas* in american english.

- The word *car* (Auto) is normally used for a motorized vehicle in british English. This word is less used in american english, where the word *automobile* is more common.
- The word *engine* (Motor) is used in british english for mechanical devices (or drives) which convert the energy of a fuel into mechanical energy. Common examples are the *internal combustion engine*, the *steam engine* and the *jet engine*. In american english, the word *motor* is more commonly used for the engine of an automobile, although words like *jet engine* and *steam engine* are used in the same way as in british english. The word engine is only applicable to *mechanical devices* and should not be used for *electric motors*.
- Abbreviations like **AC** (alternating current – Wechselstrom), **DC** (direct current – Gleichstrom) and **rpm** (revolutions per minute-Drehzahl) are sometimes differently written like for example **A.C, D.C** or **r.p.m.** However I think that the abbreviations **AC, DC** and **rpm** are the ones that are most frequently used.

I would advice all those interested in improving their English to use an English/English Dictionary. A good example is the

"Oxford Reference Dictionary"

(published by the Oxford University Press) which gives long explanations for each word, and considers many aspects of each given word like its origin, alternate forms, synonyms, etc.

Even more helpful may be one of the new electronic dictionaries on CD-ROM, like for example the

"Cambridge Advanced Learners Dictionary on CD-ROM"

(published by the Cambridge University Press)

This dictionary gives the *written word forms* in both British and American English. The reader can also (by using a series of mouse clicks) listen to the British and American *pronounciations* of the words, and can *record* his (or her) *own pronounciation* of a word, and replay it.

This dictionary is full of other helpful features, like the different meanings of words, examples of sentences using the words, common errors made by users, related words, synonyms, exercises and many other features that are too numerous to mention here. I would whole-heartedly recommend such a dictionary, because I think it is not expensive and is excellent value for money.

Teubner Lehrbücher: einfach clever

Schiehlen, Werner / Eberhard, Peter
Technische Dynamik
Modelle für Regelung und
Simulation

2., neubearb. und erg. Aufl. 2004.
XII, 251 S. (Teubner Studienbücher Technik)
Br. € 24,90
ISBN 3-519-12365-7

Wriggers/Nackenhorst/Beuermann/
Spiess/Löhnert
Technische Mechanik kompakt
Starrkörperstatik - Elastostatik -
Kinetik

2005. 515 S. Br. € 32,90
ISBN 3-519-00445-3

Pietruszka, Wolf Dieter
MATLAB in der Ingenieurpraxis
Modellbildung, Berechnung und
Simulation

2005. XII, 320 S. mit 171 Abb. u.
18 Tab. sowie 13 Beisp. Br. € 29,90
ISBN 3-519-00519-0

Sayir / Dual / Kaufmann
Ingenieurmechanik 1
Grundlagen und Statik

2004. 222 S. Br. € 19,90
ISBN 3-519-00483-6

Sayir / Dual / Kaufmann
Ingenieurmechanik 2
Deformierbare Körper

2004. 333 S. Br. € 25,90
ISBN 3-519-00484-4

Sayir / Kaufmann
Ingenieurmechanik 3
Dynamik

2005. 278 S. Br. ca. € 24,90
ISBN 3-519-00511-5

Stand Juli 2006.
Änderungen vorbehalten.
Erhältlich im Buchhandel
oder beim Verlag.

B. G. Teubner Verlag
Abraham-Lincoln-Straße 46
65189 Wiesbaden
Fax 0611.7878-400
www.teubner.de